GABA Mechanisms in Epilepsy

GABA Mechanisms in Epilepsy

Edited by

Godfrey Tunnicliff
Laboratory of Neurochemistry
Indiana University School of Medicine
Evansville, Indiana

Beat U. Raess
Department of Pharmacology and Toxicology
Indiana University School of Medicine
Evansville, Indiana

WILEY-LISS

A JOHN WILEY & SONS, INC., PUBLICATION
New York • Chichester • Brisbane • Toronto • Singapore

Library of Congress Cataloging-in-Publication Data

GABA mechanisms in epilepsy / edited by Godfrey Tunnicliff, Beat U. Raess.

    p. cm.

    Includes bibliographical references and index.

    ISBN 0–471–56827–9

    1. Epilepsy—Pathophysiology.   2. GABA—Agonists—Therapeutic use—–Testing.   3. GABA—Pathophysiology.   4. GABA—Mechanism of action.

I. Tunnicliff, Godfrey.   II. Raess, B.U.

    [DNLM: 1. Epilepsy—drug therapy.   2. Epilepsy—physiopathology.

3. GABA—antagonists & inhibitors.   4. GABA—physiology.   WL 385

G112]

RC372.5.G33   1991

616.8'53—dc20

DNLM/DLC

for Library of Congress   91–15341

                                                  CIP

Contents

Contributors

Wayne E. Crill, Department of Physiology and Biophysics, University of Washington, Seattle, WA 98195 [31]

Hiroshi Fukuzako, Department of Neuropsychiatry, Faculty of Medicine, Kagoshima University, Kagoshima 890, Japan [1]

Roger W. Horton, Department of Pharmacology and Clinical Pharmacology, St. George's Hospital Medical School, London SW17 ORE, England [121]

Kanji Izumi, Department of Pharmacology, Faculty of Medicine, Kagoshima University, Kagoshima 890, Japan [1]

Kresimir Krnjević, Anaesthesia Research Department, McGill University, Montréal, Quebec H3G 1Y6, Canada [47]

Povl Krogsgaard-Larsen, PharmaBiotec Research Center, Department of Organic Chemistry, Royal Danish School of Pharmacy, Copenhagen Ø, DK-2100, Denmark [165]

Orla M. Larsson, PharmaBiotec Research Center, Department of Biology, Royal Danish School of Pharmacy, Copenhagen Ø DK-2100, Denmark [165]

Robert L. Macdonald, Departments of Neurology and Physiology, University of Michigan Medical Center, Ann Arbor, MI 48104 [89]

Brian S. Meldrum, Department of Neurology, Institute of Psychiatry, London SE5 8AF, England [205]

Beat U. Raess, Department of Pharmacology and Toxicology, Indiana University School of Medicine, Evansville, IN 47712 [ix,105]

Arne Schousboe, PharmaBiotec Research Center, Department of Biology, Royal Danish School of Pharmacy, Copenhagen Ø DK-2100, Denmark [165]

Maharaj K. Ticku, Department of Pharmacology, The University of Texas Health Science Center at San Antonio, San Antonio, TX 78284-7764 [149]

Godfrey Tunnicliff, Laboratory of Neurochemistry, Indiana University School of Medicine, Evansville, IN 47712 [ix,105,189]

Roy E. Twyman, Department of Neurology, University of Michigan Medical Center, Ann Arbor, MI 48104 [89]

The numbers in brackets are the opening page numbers of the contributors' articles.

Preface

Clinical seizure disorders are symptoms of uncontrolled, paroxysmal hypersynchronous discharges in cerbral gray matter and are manifestations of a multitude of pathophysiological events. The basis of these events typically involves congenital, developmental, or acquired alterations in cerebral structures, unbalanced cellular metabolism, and interruption of normal cortical and subcortical integration. Aside from easily correctable causes of epileptic activitiy such as drug toxicity, hypoglycemia, and electrolyte imbalances, there remains the formidable challenge of understanding the exact etiology and therefore developing a rational basis of successful pharmacotherapy for secondary symptomatic epilepsy.

Despite our limited grasp of the mechanisms underlying epilepsy, the use of drugs to alleviate the symptoms is, to an extent, an illustration of how knowledge gained by basic research can be beneficial clinically. Even so, a significant number of patients with epilepsy (at least ten percent, representing about 200,000 patients in the United States) do not have their seizures completely controlled. Realistically, their greatest hope for seizure control is the emergence of anticonvulsant agents that are more effective than those currently available. To a large extent, though, success in the search for these drugs depends on a better comprehension of the pathophysiological processes behind the epilepsies.

An enormous amount of interest is being shown in the excitatory amino acids, and much effort is being expended in studying ways to develop anticonvulsants that interfere with excitatory amino acid neurotransmission. Nevertheless, several compelling lines of evidence point to the idea that the inhibitory capacity of the central nervous system, represented primarily by GABA-mediated neurotransmission, is altered in certain convulsive disorders. Our objective when planning this book was to bring together in one convenient source a series of authoritative articles that would review, analyze, and discuss this evidence. In so doing, we hoped this would clarify what we know of GABA's part in epilepsy and lead to a stimulation of further research in this field.

We felt it important that this series of specialized, topical reviews start with chapters that would lay the groundwork for the ensuing articles. Consequently, both the clinical aspects of the epilepsies and the role of GABA in CNS function are thoroughly discussed before the appearance of the chapters on the involvement of GABA in models of epilepsy and human epilepsy. Several other chapters review and discuss the rationale for the use of GABA-mimetics as anticonvulsants. The concluding chapter by Professor Meldrum examines the likelihood that drugs that enhance GABA function, especially by acting at the $GABA_A$ receptor, will be at the forefront of future anticonvulsant therapy.

This book will be a valuable addition to the personal library of any neuroscientist or clinician interested in understanding seizure mechanisms, particularly those at the biochemical level.

G. Tunnicliff
B. U. Raess
Evansville, Indiana

GABA Mechanisms in Epilepsy, pages 1–30
© 1991 Wiley-Liss, Inc.

1
Clinical Aspects of the Epilepsies

HIROSHI FUKUZAKO AND KANJI IZUMI

*Departments of Neuropsychiatry (H.F.) and Pharmacology (K.I.),
Faculty of Medicine, Kagoshima University, Kagoshima 890, Japan*

INTRODUCTION

Epilepsy is a neuropsychiatric disorder with relatively high prevalence and annual incidence rates. According to a considerable number of epidemiological studies from many countries its prevalence and incidence range from 0.15 to 1.95% and from 0.02 to 0.05%, respectively (Zielinski, 1988). The incidence is higher within the first decade and particularly in the first year of life. Males tend to predominate in both prevalence and incidence.

Epilepsy is not a disease but a syndrome of many different cerebral disorders. It may be defined as a disturbance of brain function of various etiologies characterized by recurrent seizures due to excessive fluctuations in cerebral electrochemical balance associated with a variety of clinical and laboratory manifestations.

Recent technological advances in combined video and electroencephalographic (EEG) monitoring have provided an accurate distinction between epileptic and nonepileptic seizures, and between various types of epileptic seizures, leading to improved treatment. In addition, recent advances in neuroradiological techniques such as computed tomography (CT), magnetic resonance imaging (MRI), and positron emission tomography (PET) have made it possible to detect some of the underlying causes of seizures and to identify the epileptogenic focus without invasive methods. In this chapter we shall describe the classification, diagnosis, and treatment of epilepsy with reference to these recent advances. We shall also discuss at the end of the chapter a postural myoclonus that we have recently found to be associated with the long-term administration of neuroleptics.

CLASSIFICATION

The classification of epilepsy is complex, and can be based on the etiology, pathology, age of onset, clinical seizure, EEG findings, or prognosis. Epilepsies have long been divided for convenience into two groups—those caused by recognizable cerebral lesions and those in which there was no identifiable cause. A revised classification of individual seizure types was accepted in

1981 by the General Assembly of the International League Against Epilepsy (ILAE) and has been widely used in the management of epilepsy (Commission on Classification and Terminology of the ILAE, 1981) (Table I). However, this does not accurately reflect the complicated epileptic syndrome which is characterized by a cluster of signs and symptoms customarily occurring together. These include such items as type of seizure, etiology, anatomy, precipitating factors, age of onset, chronicity, diurnal and circadian cycling, and sometimes prognosis. The classification of epilepsies and epileptic syndromes was therefore devised in 1985. Recently, a new classification of epilepsies and epileptic syndromes has been published by the Commission on Classification and Terminology of the ILAE (1989) (Table II). The principles of division are whether the seizure is localization-related or generalized and whether the discernible etiology of seizure is present, not present, or not known: symptomatic, idiopathic, or cryptogenic. Using these classifying principles, epileptic syndromes are now classified as in Table II. This classification is not totally satisfactory and is still undergoing evaluation, although several of the epileptic syndromes are quite well

TABLE I. International Classification of Epileptic Seizures in 1981

   I. Partial seizures
      A. Simple partial seizures
         1. With motor symptoms
         2. With somatosensory or special sensory symptoms
         3. With autonomic symptoms
         4. With psychic symptoms
      B. Complex partial seizures
         1. Simple partial onset followed by impairment of consciousness
            a. With no other features
            b. With features as in A.1–4
            c. With automatisms
         2. With impairment of consciousness at onset
            a. With no other features
            b. With features as in A.1–4
            c. With automatisms
      C. Partial seizures evolving to secondarily generalized seizures
  II. Generalized seizures
      A. 1. Absence seizures
         2. Atypical absence seizures
      B. Myoclonic seizures
      C. Clonic seizures
      D. Tonic seizures
      E. Tonic-clonic seizures
      F. Atonic seizures
 III. Unclassified epileptic seizures

Modified from Commission on Classification and Terminology of the International League Against Epilepsy.

**TABLE II. International Classification
of Epilepsies and Epileptic Syndromes in 1989**

I. Localization-related (focal, local, partial) epilepsies and syndromes
    A. Idiopathic (with age-related onset)
        1. Benign childhood epilepsy with centrotemporal spike
        2. Childhood epilepsy with occipital paroxysms
        3. Primary reading epilepsy
    B. Symptomatic
        1. Chronic progressive epilepsia partialis continua of childhood
           (Kojewnikow's syndrome)
        2. Syndromes characterized by seizures with specific modes of
           precipitation
        3. Temporal lobe epilepsies
        4. Frontal lobe epilepsies
        5. Parietal lobe epilepsies
        6. Occipital lobe epilepsies
    C. Cryptogenic
II. Generalized epilepsies and syndromes
    A. Idiopathic (with age-related onset-listed in order of age)
        1. Benign neonatal familial convulsions
        2. Benign neonatal convulsions
        3. Benign myoclonic epilepsy in infancy
        4. Childhood absence epilepsy (pyknolepsy)
        5. Juvenile absence epilepsy
        6. Juvenile myoclonic epilepsy (impulsive petit mal)
        7. Epilepsy with grand mal (GTCS) seizures on awakening
        8. Other generalized idiopathic epilepsies not defined above
        9. Epilepsies with seizures precipitated by specific modes of activation
    B. Cryptogenic or symptomatic (in order of age)
        1. West syndrome (infantile spasms, Blitz-Nick-Salaam Krämpfe)
        2. Lennox-Gastaut syndrome
        3. Epilepsy with myoclonic-static seizures
        4. Epilepsy with myoclonic absences
    C. Symptomatic
        1. Non-specific etiology
           a. Early myoclonic encephalopathy
           b. Early infantile epileptic encephalopathy with suppression burst
           c. Other symptomatic generalized epilepsies not defined above
        2. Specific syndromes
III. Epilepsies and syndromes undetermined whether focal or generalized
    A. With both generalized and focal seizures
        1. Neonatal seizures
        2. Severe myoclonic epilepsy in infancy
        3. Epilepsy with continuous spike-waves during slow wave sleep
        4. Acquired epileptic aphasia (Landau-Kleffner-syndrome)
        5. Other undetermined epilepsies not defined above
    B. Without unequivocal generalized or focal features

(Continued)

**TABLE II. International Classification
of Epilepsies and Epileptic Syndromes in 1989 (Continued)**

IV. Special syndromes
   A. Situation-related seizures (Gelegenheitsanfälle)
      1. Febrile convulsions
      2. Isolated seizures or isolated status epilepticus
      3. Seizures occurring only when there is an acute metabolic or toxic
         event due to factors such as alcohol, drugs, eclampsia, nonketotic
         hyperglycemia

Modified from Commission on Classification and Terminology of the International
League Against Epilepsy.

defined. Nevertheless, this classification is probably of greater benefit than
one based on simple seizure diagnosis.

The 1981 classification of epileptic seizures (Table I) is considered pragmatic for clinical use, though it is somewhat complicated for clinicians except for specialists in epilepsy. Classification of epileptic seizures serves as a useful tool for guiding decisions about when to treat epilepsy and how to choose among the available antiepileptic drugs (Commission on Classification and Terminology of the ILAE, 1981). The classification is based on the clinical event and its accompanying EEG. Epileptic seizures are conventionally divided into two groups: partial or generalized. The distribution of seizure types varies between prospective and retrospective patient reports. Moreover, the patients vary in age and are from institutions of divergent characteristics. In recent papers on seizure types in adults, almost similar results have been reported as demonstrated in Table III (Keränen et al.,

**TABLE III. Distribution of Seizure Types
in 1,005 Epileptic Patients With Classifiable Seizures**

Seizure types	No. of cases	%
Partial seizures	682	67.9
Simple partial	92	9.2
Complex partial	280	27.9
Partial seizures secondarily generalized	310	30.8
Generalized seizures	323	32.1
Absence seizures	17	1.7
Myoclonic seizures	13	1.3
Tonic or clonic seizures	7	0.7
Tonic-clonic seizures	283	28.1
Atonic seizures	3	0.3
Total	1,005	100.0

Modified from Keränen et al. (1988).

1988). These authors show that partial seizures predominate over generalized seizures and that tonic-clonic seizure is the most frequent type of generalized seizure. The signs and symptoms of the chief types of seizure as presented in Table I are examined below.

Partial Seizures

Clinical and EEG findings show that partial seizures arise from unilateral brain structures. Partial seizures can be divided into three types according to whether consciousness is preserved and whether a seizure is generalized or not. Simple partial seizures do not impair consciousness whereas complex partial seizures do. Preservation of consciousness is generally judged on the basis of the ability to respond purposefully and to retain memory. In secondarily generalized partial seizures there is a spreading of local discharge and an evolvement into generalized seizures.

Simple partial seizures.   Symptoms often reflect the region in which the seizures originate. The seizures are often followed by complex partial or secondarily generalized tonic-clonic seizures. Almost all seizures end within 3 min (Devinsky et al., 1988). Simple partial seizures encompass a diverse clinical spectrum of motor, sensory, cognitive, and psychic signs and symptoms, and can be divided into four types as follows:

1. Motor symptoms include various movements of the body known as focal motor seizure, jacksonian seizure, adversive seizure, versive seizure, and vocal seizure.

2. Somatosensory symptoms are usually composed of feelings of numbness, deadness, pins and needles, and sometimes pain or burning.

3. Special sensory symptoms are described as visual, auditory, olfactory, gustatory, and vertigeous feelings such as double vision, microscopia, hallucination, and unpleasant odors and tastes.

4. Autonomic symptoms include headache, vertigo, nausea, vomiting, abdominal pain, dyspnea, and perspiration. These symptoms are observed as single manifestations or in combined forms. Psychic symptoms almost always occur as part of complex partial seizures or secondarily generalized tonic-clonic seizures. During the seizure, the patient may experience disorders of thought, emotion, and perception. Symptoms involving thought are described as feelings of familiarity or unreality e.g., déja vu or jamais vu. Emotional experiences commonly contain intense fear, sometimes laughter or depression, and very occasionally pleasure or ecstasy. Perceptual disturbance, which includes visual distortions of size, distance, or shape of the object, distortion of sound, visual and auditory hallucinations, and hallucinations of taste and smell, may occur.

Complex partial seizures.   Complex partial seizures are typically characterized by automatisms. Auras may include any of the symptoms of simple partial seizures, especially vague feelings of discomfort. It has been considered that complex partial seizures begin with auras. Recent observations, however, using videotape techniques suggest that auras in complex partial seizures are not as frequent as was previously believed (Theodore et al., 1983). On the average the seizures last for about 3 min including a postictal period.

Automatism is the clouding of consciousness during which the individual retains control of posture and muscle tone but performs simple or complex movements and actions without being aware of what is happening (Fenton, 1972). Most automatisms are simple, stereotypic, and aimless movements. Simple automatisms include swallowing, chewing, lip smacking, tapping, and random eye movements. They often seem to mimic what patients have been doing before the beginning of the attack. More elaborate automatisms include well-coordinated but inappropriate undressing, drinking, wandering, and running that can be mistaken for normal actions. Violent behavior may occur but this is rare. Most automatisms are associated with epilepsy originating in the temporal lobe. Complex partial seizures are apt to be confused with absence seizures. Complex partial seizures can be distinguished from absence seizures by the clinical features of patients: absence seizures present no postictal confusion, last less than 10 sec in most cases, and the affected body areas are usually restricted to the face and arms (Theodore et al., 1983).

Generalized Seizures

Absence seizures. Absence seizures consist almost solely of brief lapses of consciousness that last more than 3 sec. They begin and end abruptly and recur from a few to several hundreds times per day. Phenomenologically, the patient suddenly stops moving or shows automatic behavior such as inconspicuous flickering of the eyelids, rubbing of the nose, chewing, or swallowing. Falling does not occur. The patient immediately regains consciousness after the lapse. Intelligence is not usually impaired but sometimes deteriorates in patients who also have other types of seizures (Fois et al., 1987).

Myoclonic seizures. Myoclonic seizures are bilaterally synchronous, shock-like jerks of limb and truncal muscles. Various disorders including progressive cerebral degeneration, head trauma, cerebral anoxia, and drug intoxication cause such seizures.

Clonic seizures. Clonic seizures are occasionally seen when a generalized tonic-clonic seizure lacks a tonic phase. The postictal phase of this seizure is shorter than in tonic-clonic seizure.

Tonic seizures. Tonic seizures represent a rigorous symmetrical tonic contraction of muscles. The patient maintains a posture such as holding both arms over the head. The duration of this seizure is usually brief (5–10 sec) and is not necessarily accompanied by loss of consciousness. Tonic seizures are a necessary component of the Lennox-Gastaut syndrome (Beaumanoir, 1985).

Tonic-clonic seizures. Tonic-clonic seizures are observed in both generalized and secondarily generalized seizures. Generalized tonic-clonic seizure is said to occur when the clinical findings and EEG indicate bilateral simultaneous involvement at the onset of the seizure. In secondarily generalized tonic-clonic seizure the clinical findings and EEG indicate that the seizure begins unilaterally. Tonic-clonic convulsions are characterized by loss of consciousness and falling. As the patient falls, the body stiffens because of generalized tonic contraction of the axial limb muscles. During this tonic phase, which lasts less than a minute, respiration stops and pallor or cyanosis may be seen. After the tonic stage, clonic movements occur in the extremities for less than a minute. The tongue is sometimes bitten and frothing at the mouth is occasionally observed. Urinary incontinence may be seen. A period of more relaxed unconsciousness follows and lasts for about a minute. Afterward, the patient is confused, sleepy, and uncooperative for several minutes before full recovery. The patient does not remember what happened during the seizure or even, in most cases, the fact that a seizure has occurred.

Atonic seizures.   Atonic seizures are characterized by abrupt loss of muscle tone, often causing the patient to fall and suffer injury. They usually start in childhood and persist into adult life. Mental retardation with abnormal neurological findings is often present. Atonic seizures are notoriously difficult to treat with antiepileptic drugs. These seizures are often associated with the Lennox-Gastaut syndrome that is characterized by impaired consciousness, atypical absences, and slow spike-wave discharges (Fig. 1).

Other Types of Seizures

The following epilepsies can be seen in Table II, but not in Table I.

Benign partial epilepsy with centrotemporal spikes.   Benign epilepsy of childhood has a very characteristic clinical feature of predominantly nocturnal partial seizures between 3 and 13 years of age, and shows an impressive EEG (Lerman, 1985). Seizures usually have a brief duration, perhaps 2 min at the most. The seizures are characterized by a somatosensory onset with unilateral paresthesias involving the tongue, lips, gums, and inner cheeks; followed by unilateral tonic, clonic, or tonic-clonic convulsions which involve the face, lips, tongue, pharynx, and larynx, and lead to speech arrest or anarthria and drooling. Diurnal seizures never become generalized and consciousness is preserved. Typically, spikes or spike-wave discharges of high amplitude are localized in the unilateral or bilateral centrotemporal region with normal background activity (Fig. 2). The centrotemporal spikes increase with drowsiness and during sleep. This epilepsy is highly sensitive to antiepileptic drugs and disappears by the age of 20 even without treatment.

Neonatal seizures.   Seizures in the newborn are likely to be symptomatic of epilepsy in later childhood. The seizures can take a number of clinical forms. Neonatal seizures are classified into 4 types: subtle, clonic, tonic, and myoclonic. Clonic seizures are the most common type in newborn infants, but tonic seizures are more common in premature infants and occur particularly in association with intraventricular hemorrhage. Myoclonic seizures are considered to be very rare in the neonatal period. The common causes of neonatal seizures are congenital malformations, hypoxia, infections, trauma, intracranial hemorrhage, and various metabolic disorders such as hypoglycemia, hypocalcemia, hypomagnesemia, hypo- and

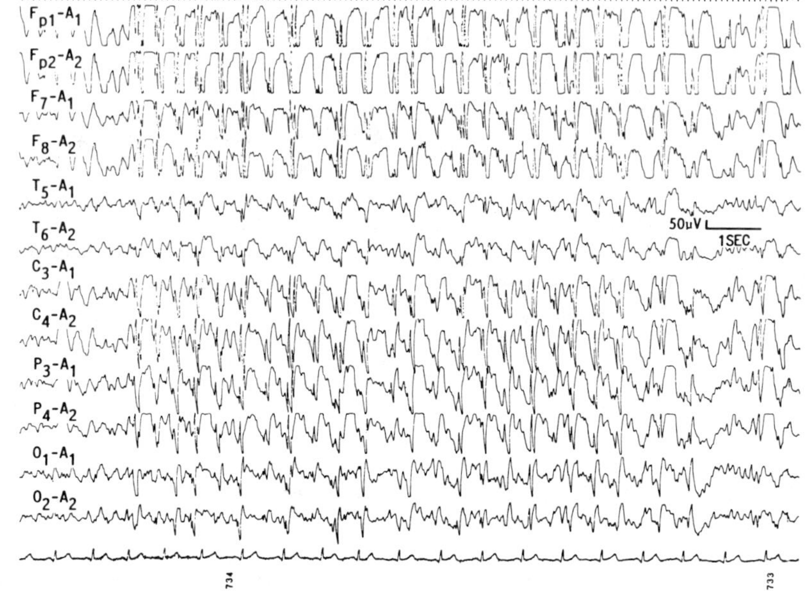

1

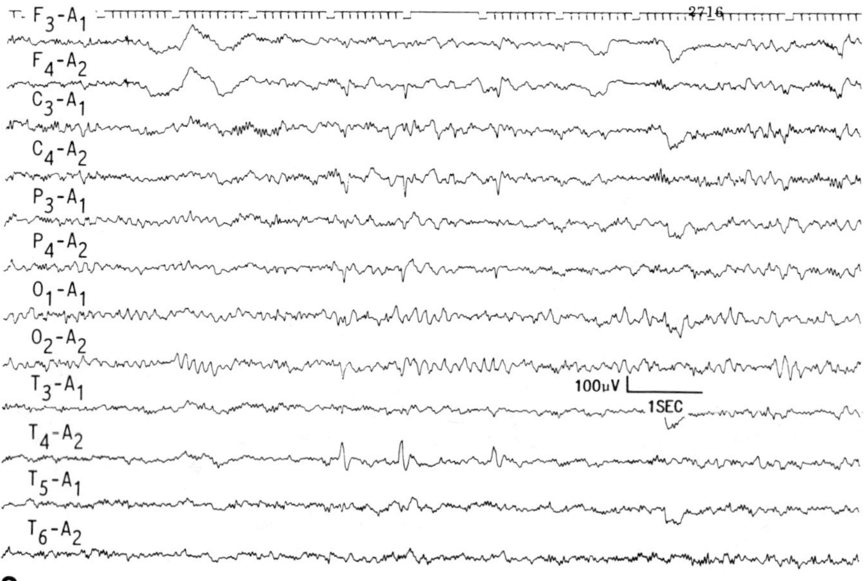

2

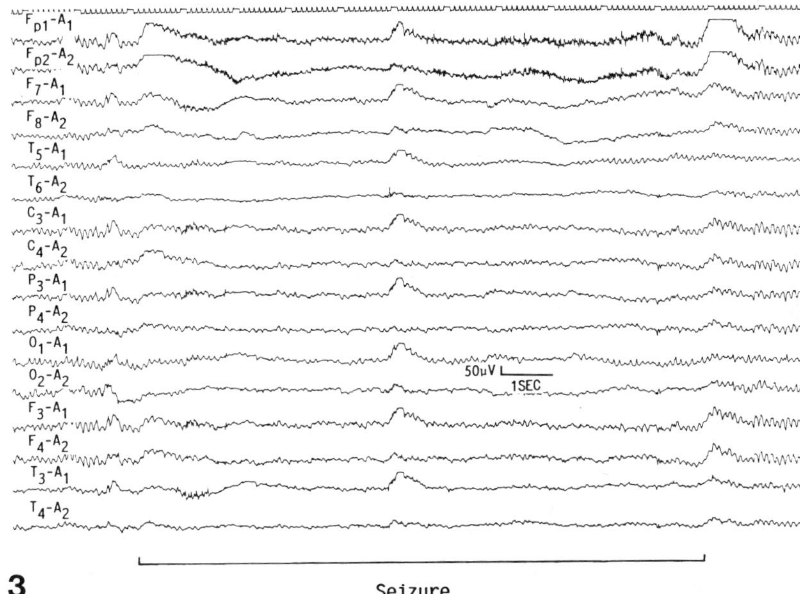

Fig. 3.   An EEG during tonic seizure which is suddenly precipitated when the patient stamps on the floor. The EEG shows attenuation and fast activity during the seizure followed by restoration of alpha activity.

hypernatremia, hyperammonemia, pyridoxine deficiency, and disorders of amino acid metabolism (Engel, 1989).

The EEG is an important tool in both the diagnosis and management of neonatal seizures. Simultaneous clinical and EEG observations may be necessary to confirm the diagnosis of subtle seizures. Successful therapy may only be possible by the use of continuous EEG monitoring. The prognosis of newborn babies who have seizures depends on the underlying etiology. The earlier the seizures occur, the greater the probability of later cerebral palsy, mental retardation, and epilepsy (Holden et al., 1982). Neonates who have seizures should be kept in an oxygen tent and their general biochemical status monitored. In the absence of a remediable metabolic disorder, neonatal seizures may be treated with antiepileptic

---

Fig. 1.   An EEG of a patient with the Lennox-Gastaut syndrome. Slow polyspike-wave discharges are seen during atypical absence seizures.

Fig. 2.   An EEG of a patient with benign partial epilepsy with centrotemporal spikes. Spikes or sharp waves mainly emerge in the right centrotemporal region.

drugs. Phenobarbital is usually the first choice. Surgery may be the best treatment of subdural hematoma secondary to birth injury.

Febrile convulsions.   Febrile convulsions occur in children with predisposing factors only when a feverish illness occurs during the critical age period. Febrile convulsions occur almost exclusively before the age of 5 years with an incidence of about 3%. The main clinical significance of febrile convulsions is that the risk of later epilepsy in these children is 5 times that in children who did not convulse when febrile. Risk factors for the later development of epilepsy are preexisting neurological abnormalities or mental retardation, prolonged febrile convulsions, focal features, family history of afebrile seizures, and increasing frequency of febrile convulsions (Nelson and Ellenberg, 1978; Annegers et al., 1979). EEG offers little useful information on the prognosis following febrile convulsions. In children with some risk factors for recurrence, prophylactic antiepileptic medication is needed. Valproate may be appropriate because it has few effects on cognitive function and behavior (Wallace, 1981).

Reflex epilepsy.   Reflex epilepsy occurs in response to a fixed and clearly recognized sensation or perception such as flickering light, noise, music, touch, reading, and movement. These stimuli are usually limited in individual patients to a single specific stimulus or a small number of closely related stimuli. Photosensitive epilepsy is the most common form of such epilepsy. Although generalized seizures are commonly seen, partial seizures may also be precipitated by proprioceptive stimuli. Figure 3 shows an EEG during tonic seizure provoked by sudden movement of the patient.

Status Epilepticus

Status epilepticus consists of recurrent seizures without recovery of consciousness between attacks. Three types are known: convulsive status, nonconvulsive status, and continuous partial seizures. Convulsive status occurs in about 3% of epileptic patients. In patients chronically treated with antiepileptic drugs, sudden cessation of drugs is a common precipitating factor. In the case of status as an initial symptom it is important to find the recognizable causes. The causes include brain tumor, cerebral thrombosis, meningitis, and metabolic abnormality. Prolonged seizure activity by itself may damage the brain. Excessive continuing neuronal firing, therefore, must be stopped as soon as possible. Nonconvulsive status consists of generalized nonconvulsive (absence) and complex partial statuses. In contrast to typical absence status featuring solely decreased consciousness, atypical absence status shows not only decreased consciousness but also associated phenomena such as myoclonus and automatisms. There may be cases of complex partial status showing pure mental confusion. These two statuses are therefore very difficult to distinguish (Tomson et al., 1986). Continuous partial seizures include any forms of simple partial status. There are no well-documented statistics on the incidence of this status. The most common form is the continuous clonic focal motor seizures i.e., Kojewnikow's syndrome.

DIAGNOSTIC EVALUATION OF THE PATIENT

The most important component in evaluating a seizure disorder is a detailed history of the patient. Observation of the seizure could confirm the diagnosis of epilepsy but this would be unusual; otherwise diagnosis would depend on the description of epileptic seizures from witnesses. Seizures in infancy and early childhood would also be important for diagnosis. In addition, the use of drugs and alcohol must be investigated in detail. Even if the diagnosis of epilepsy is confirmed from the history and physical examination of the patient, further examination is usually needed. This examination should include a complete blood count, urinalysis, electrolytes, calcium, blood sugar, liver function, serum copper, and a test for syphilis. The cause of the seizures is rarely detected by these blood tests, but they provide baselines for subsequent monitoring of the chronic effects of drug treatment.

Electroencephalography

EEG plays an important role in both the diagnosis and treatment of epilepsy. Electrodes are placed on the scalp according to the international placement or the 10–20 system. Nasopharyngeal and sphenoidal electrodes are often valuable in revealing abnormal activity in the anterior- and medial-temporal areas. In normal adults with the eyes closed, two frequency bands are usually seen: alpha activity at 8 to 13 Hz in the posterior regions and beta activity at 14 to 35 Hz more anteriorly. Continuous 25 Hz activity is seen in the central regions and 17 to 20 Hz activity mixed with 6 to 8 Hz waves is found in the intermediate frontal areas. In approximately 50% of patients with epilepsy, epileptiform activity is demonstrated with a single waking EEG. Repeated EEGs reveal it in 85% of epilepsy cases (Ajmone-Marsan and Zivin, 1970). Some EEG rhythms of uncertain clinical significance may be confused with epileptiform activities. For example, brief bursts of positive spikes at around 6 and 14 Hz are often mistaken for epileptiform discharges. They are most evident in the posterior area unilaterally or bilaterally during drowsiness and light sleep. Runs of 4 to 7 Hz phantom spike-wave activity are seen diffusely with frontal or occipital predominance. The latter is not evidence for epilepsy (Hughes, 1980). Bursts of 4 to 7 Hz rhythmic temporal discharges lasting for a few seconds to several minutes may be seen unilaterally or bilaterally. This activity is currently considered to be of uncertain clinical significance. Small sharp spikes, wicket spikes, and breach rhythm have no clinical significance and must be differentiated from genuine epileptogenic discharges. To increase epileptiform activity, activation procedures such as hyperventilation, photic stimulation, sleep induction and deprivation, and convulsant administration are used (Engel, 1989). Figure 4 represents a case of myoclonic epilepsy with spikes or spike-wave discharges provoked by photic stimulation.

The major clinical demand for long-term EEG monitoring arises when the diagnosis of epilepsy is doubtful. Demonstration of appropriate ictal EEG changes can establish the diagnosis of epilepsy (Dreifuss, 1986).

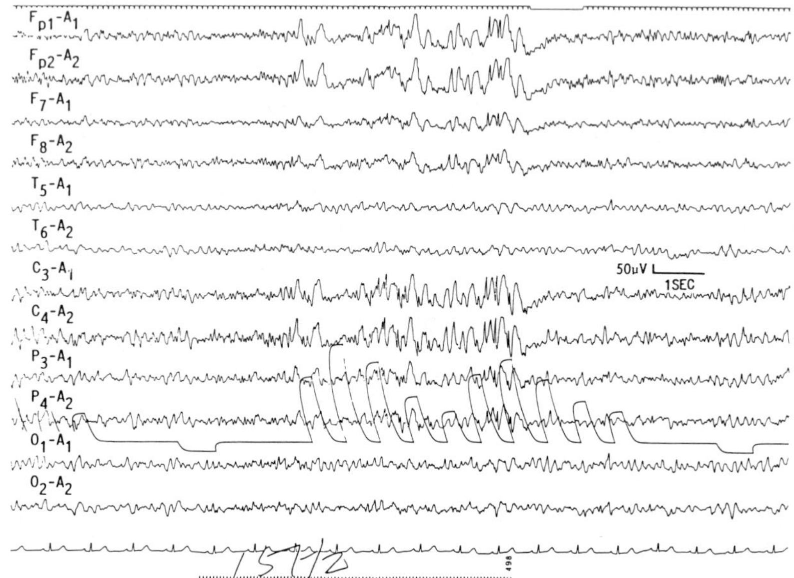

4

5

Long-term EEG together with video recording of behavior is necessary for the accurate recognition of clinical or electrical events during seizures, and is useful in classification, with possible implications for future treatment. Monitoring may also be valuable to determine the frequency of seizures that are not reliably detected by the patient or other witnesses. Monitoring may even be required to determine whether known EEG discharges are accompanied by clinical ictal events. If necessary, continuous psychological tests are employed; these often demonstrate transitory cognitive impairment during apparently subclinical discharges (Aarts et al., 1984).

Simple partial seizures. Simple partial seizures usually reveal no ictal change on the scalp EEG. Ictal EEG abnormalities are more commonly observed in motor than in non-motor simple partial seizures (Devinsky et al., 1988). In interictal EEG, a paroxysmal discharge of spikes, sharp waves, or theta activity may be observed, though not necessarily focal or even unilateral. Before these abnormalities depth EEG exhibits rapid and spiky discharges, appearing in limbic structures in many cases. Epileptiform interictal activity often accompanies simple partial seizures in patients who have complex partial or secondarily generalized tonic-clonic seizures. On the other hand, the frequency of interictal changes in patients who only have simple partial seizures is probably low. In addition to epileptiform activity, some patients display local slowing, generalized slowing, or intermittent slow wave activity. Devinsky et al. (1989) recommend subdural electrode recordings to localize the onset and spread of simple partial seizures even if the seizure originates from the medial temporal lobe.

Complex partial seizures. The pattern of ictal EEG varies both within and between subjects. However, a focal or generalized reduction in amplitude of ongoing activity is often seen as an initial EEG change. During complex partial seizures the EEG shows intermittent or continuous localized epileptiform discharges, diffuse slow activity, or bilateral or multifocal discharges. The interictal EEG shows spikes, sharp waves, or diffuse slow activity, occasionally normal. Focal epileptiform activity occurs most commonly in the temporal regions (Fig. 5), and less commonly in the frontal or occipital region. Many patients with complex partial seizures display more than one focus. Most often the foci are in homologous regions of the right and left hemispheres and sometimes they may occur synchronously or more often with a small delay between one side and the other. Postictally, the EEG may show localized or diffuse reduction in amplitude, followed by slow

---

Fig. 4. An EEG of a patient with myoclonic epilepsy showing photosensitivity. Photic stimulation precipitates spikes and spike-wave discharges.

Fig. 5. An EEG of a patient with complex partial seizures. Polarity of spikes inverts at left anterior- and centrotemporal regions. This phenomenon indicates that there are epileptic foci in these regions.

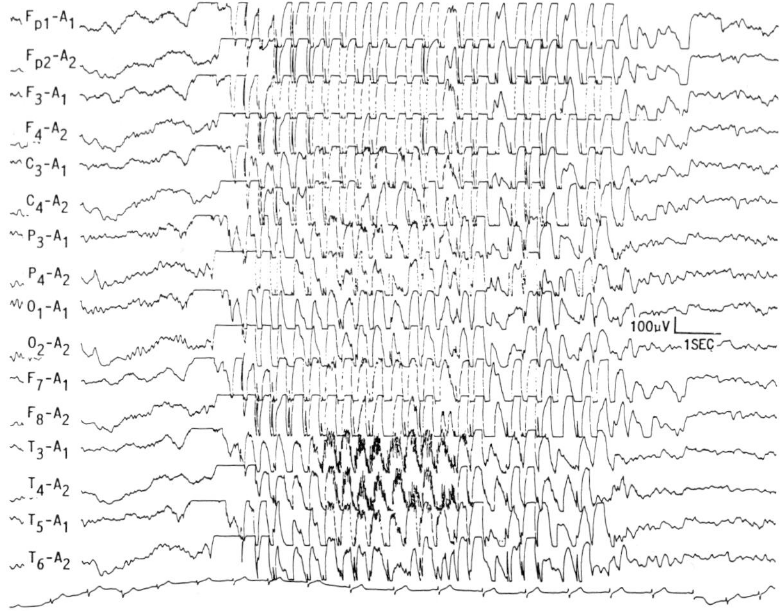

6

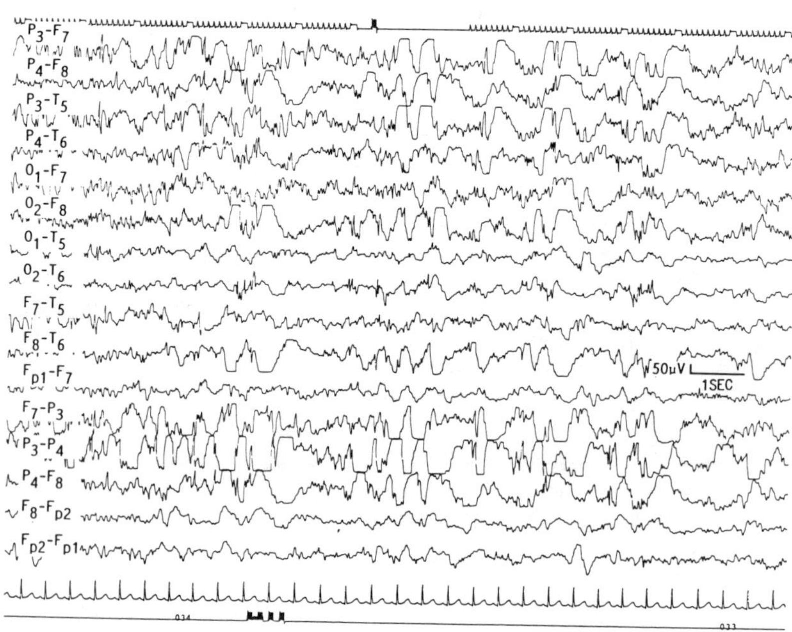

7

waves, either localized or generalized. Persistance of the postictal abnormality may vary from a few minutes to a few days.

Absence seizures.   The EEG during typical absence seizures is characterized by bursts of generalized, fairly symmetrical 3 Hz spike-wave discharges which commence abruptly, usually 1 or 2 sec before the onset of apparent clinical manifestations, and may continue briefly after consciousness has begun to return (Fig. 6). They usually last about 10 sec and cease abruptly. Postictal changes are not seen. Bursts of bilateral or generalized spike-wave discharges at 1.5 to 2.5 Hz characterize atypical absence seizures. Their onset and end are more gradual than with typical absence seizures. Absence seizures with prominent myoclonic jerks are often accompanied by polyspike-wave discharges.

Myoclonic seizures.   The EEG shows multiple spikes or polyspike-wave, spike-wave, or sharp-wave discharges. Similar findings are seen on interictal EEG.

Clonic seizures.   Generalized fast activity of about 10 Hz or more is mixed with spike-wave or polyspike-wave discharges during seizure. Interictal EEG also exhibits spike-wave or polyspike-wave discharges.

Tonic seizures.   The EEG displays low voltage, fast activity of 9 to 10 Hz or more, which decreases in frequency and increases in amplitude.

Tonic-clonic seizures.   With tonic-clonic convulsion a brief electrodecremental event signals the onset of generalized seizure. This is followed by a burst of spikes and then by generalized spike-wave activity. In the tonic phase, the tracing is usually obscured by muscle artefact. In the clonic phase, muscle jerks emerge synchronously with the spike-wave discharges and, as the seizure ends, both become less frequent and irregular. In the postictal state, generalized slowing is seen, possibly preceded by a period of low amplitude. Interictal 3.5 to 4.5 Hz generalized spike-wave discharges are common in patients who have this seizure. Photic stimulation often precipitates bilateral paroxysmal discharges in patients who have primary generalized epilepsy. Hyperventilation provoked EEG epileptic abnormalities in about 40% of patients with primary generalized epilepsy and in 16% of those with secondarily generalized epilepsy (D'Alessandro et al, 1986).

Atonic seizures.   Ictal changes in atonic seizures may manifest as generalized spike-waves, polyspike-waves, slow spike-wave discharges, or low voltage fast activity. The EEG may show only an electrodecremental event with loss of muscle tone (Egli et al., 1985).

Infantile spasms.   Typical West syndrome is constituted by spasms, mental retardation, and hypsarrhythmia on the EEG (Fig. 7). Hypsarrhyth-

---

Fig. 6. An EEG of a patient with typical absence seizures. Fairly symmetrical spike-wave discharges of 3 Hz appear during the seizure.

Fig. 7. An EEG of a patient with West syndrome. Hypsarrhythmia on EEG characterizes this syndrome.

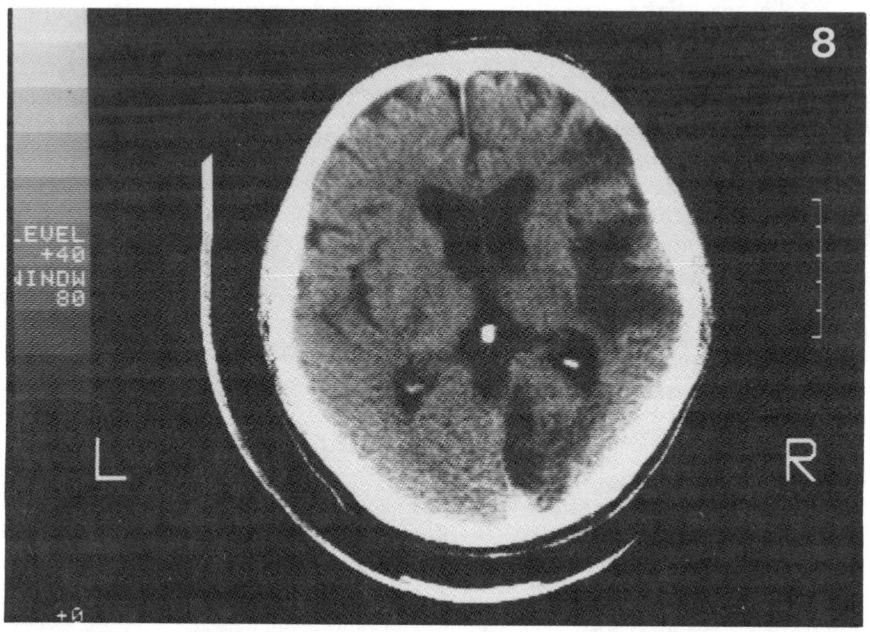

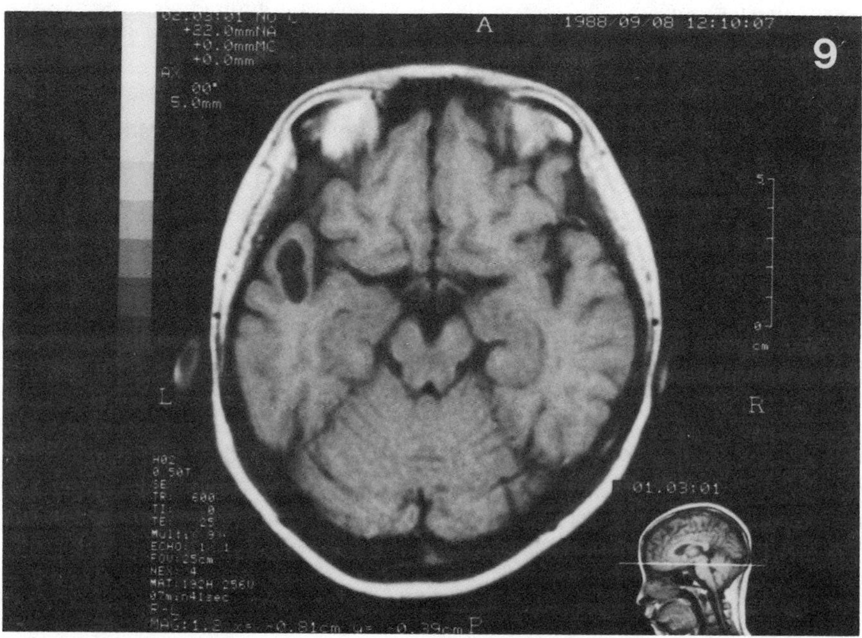

mia is defined by Jeavons (1985) as a grossly chaotic mixture of very high amplitude (more than 200μ) slow waves at frequencies of 1 to 7 Hz with spikes and sharp waves which vary in amplitude, morphology, duration, and site. This pattern is observed in two-thirds of patients with infantile spasms. There is a remarkable reduction in hypsarrhythmic recording during sleep.

Status epilepticus.   During the early periods of status epilepticus, ictal discharges appropriate to the seizure type are repeatedly seen with incomplete recovery of the postictal disturbance between seizures. Ictal EEG patterns often become obscure as the status becomes prolonged. EEG monitoring especially in the initial phase of the status has great importance in the diagnosis and treatment of status of absence and complex partial seizures (Tomson et al., 1986).

Neuroradiological Techniques

Neuroradiological methods are useful in detecting the causes of seizures especially neurosurgically correctable seizures, and also in detecting the site of the epileptic focus.

Skull X-ray.   Plain skull X-rays usually do not elucidate the causes of seizure. However, they sometimes reveal evidence of raised intracranial pressure, calcification in tumor and other diseases, localized bone thickness and erosion, and pineal shift indicative of a mass lesion.

Computed tomography.   CT has become a routine method of identifying underlying causes of epileptic seizures, such as brain tumor, head injury, cerebral abscess, and cerebrovascular disease. Young et al. (1982) recommend that CT should be reserved for patients with focal seizures, focal signs, or focal EEG abnormalities. The incidence of CT abnormalities depends on various factors especially the selection of patients. In 4 reports, the rate of CT abnormalities ranges from 24 to 63% (McGahan et al., 1979; Young et al., 1982; Ramirez-Lassepas et al., 1984; Daras et al., 1987). There was a correlation of CT abnormality with increasing age. The incidence of abnormal CT is higher in simple partial seizures and secondarily generalized partial seizures than in other types of seizures. The most common abnormality is cerebral atrophy, particularly generalized atrophy. Patients who have suffered episodes of status epilepticus have a higher incidence of cerebral atrophy. Cerebellar atrophy is occasionally associated with prolonged therapy with phenytoin. The incidence of other abnormal CT findings varies between reports. In adulthood, however, cerebrovascular diseases are the most com-

---

Fig. 8.   A CT of a patient with convulsive seizures produced by cerebral infarction. A low density area can be seen in the right hemisphere.

Fig. 9.   An MRI of a patient with grade III astrocytoma in the left superior temporal gyrus. The MRI reveals an area of decreased signal on the T1-weighted image in the left anterotemporal lobe which was not detected by CT 2 weeks before the MRI. A neurosurgical operation relieved her of seizures.

mon causes of epilepsy (Ramirez-Lassepas et al., 1984; Lühdorf et al., 1986). Figure 8 shows a CT of a patient with convulsive seizures induced by cerebral infarction. The emergence of seizures due to brain tumor is influenced by the location and characteristics of the tumor. They may be identified using both plain and contrast enhanced CT. Epileptic seizures are likely to be induced by meningioma, astrocytoma, oligodendroglioma, and metastatic brain tumor at the temporal, anterior, or parietal lobe.

Magnetic resonance imaging.   MRI has some advantages over CT. Spatial resolution has recently improved to better than 1 mm. More information about structural brain pathology can be obtained by MRI than by CT. MRI demonstrates much better the contrast between white and gray matter. The technique seems to have no biological hazards. Furthermore, no bone-induced artefacts appear. It is expected that this superiority will be seen particularly in the brain areas adjacent to the skull, such as in the middle and posterior cranial fossa. There are several reports that MRI has revealed abnormalities such as neoplasms and abnormal blood vessels in patients with partial seizures who had normal CT (Sperling et al., 1986; Theodore et al., 1986; Avrahami et al., 1987; Bergen et al., 1989) (Fig. 9). It seems therefore that MRI should be conducted in epileptic patients who have focal activity and especially in those who have intractable epilepsy even if the CT is normal.

MRI has a high sensitivity but its specificity is low. For example, an increase in the water content generally produces prolongation of the T1 and T2 relaxation times regardless of whether it is caused by edema, infarction, encephalitis, or demyelination. Calcifications are better revealed by CT than by MRI.

Positron emission tomography.   PET is an expensive technique requiring a cyclotron which produces radionucleides such as ^{18}F, ^{15}O, ^{13}N, and ^{11}C. PET is able to demonstrate not only cerebral blood flow but also regional metabolism and oxygen utilization. PET provides important information on the epileptogenic focus and has been demonstrated to be highly accurate (86%) in identifying these focal sites (Engel et al., 1982b). During an interictal period, both metabolic activity and blood flow have been shown to be reduced at epileptogenic foci (Engel et al., 1982a,b). During an ictus, marked increases in both metabolism and blood flow are seen at the focus and in sites of the subsequent spread throughout the brain (Engel et al., 1983). Since this propagated neuronal activity spreads beyond the site of the seizure origin, ictal scans are less useful in predicting the epileptogenic focus than those obtained in the interictal period (Engel et al., 1983). PET may be helpful in determining the localization of the focus and whether the seizure is partial or generalized when the EEG findings are ambigious or inconsistent. In studies of glucose utilization, approximately 70% of patients who were candidates for neurosurgery as a result of epileptic seizures have interictal zones of hypometabolism. Approximately one-third to one-half of patients who would formerly have required depth EEG before neurosurgery can bypass this invasive step because of the PET information in their presurgical evaluation (Mazziotta, 1988).

TABLE IV. Differentiation of Episodic Symptoms From Epilepsy

Conditions to be differentiated from epilepsy	Useful findings in differentiation from epilepsy
Syncope	Premonitory symptoms such as light headedness, slowed response times, fading of sounds and sight are noticed. Automatisms, typical tonic-clonic sequence, incontinence, and a true postictal phase are not usually noted. Polygraphic recordings including cardiac monitoring are useful in diagnosis.
Pseudoseizure	There is often a personal and family history of psychiatric disorder. The onset of seizure is often gradual, and repeated EEG does not usually reveal any abnormalities. Consciousness is usually preserved. The seizure is not affected by antiepileptic drugs.
Transient ischemic attack	Paresthesias due to cerebral ischemia usually do not show sequential spread from one body area to another, clonic motor activity, or loss of consciousness.
Migraine	Family history of migraine and recurrent headaches are revealed by taking a careful history despite a predominant symptom of confusional state. EEG during an occurrence of symptoms is unrevealing.
Parasomnias	Most symptoms occur during the first one-third of the night. Complete amnesia and normal EEG are observed.
Automatic behavior	Automatic behavior seen in narcoleptic patients is often confused with automatisms due to complex partial seizures. Automatic behavior shows solely amnestic episodes that are rare in epilepsy. Various stimuli that do not affect epilepsy bring the automatic behavior to a halt. Postictal periods are often in epilepsy but never in automatic behavior. Polygraphic recordings reveal short-duration stage I or REM sleep.

(Continued)

TABLE IV. Differentiation of Episodic Symptoms From Epilepsy (Continued)

Conditions to be differentiated from epilepsy	Useful findings in differentiation from epilepsy
Transient global amnesia	In contrast to epilepsy transient global amnesia almost always occurs between the fifth and eighth decades, solely shows loss of memory, and lacks prior seizures and alterations in consciousness.
Breath-holding spell	Apnea occurs as a reaction to noxious stimuli producing anger and frustration. Neurological examination and EEG is unrevealing.
Pallid infantile syncope	Pallor and unconsciousness are usually precipitated by fright or head blow. Ocular compression leads to severe bradycardia or asystole associated with EEG slowing and loss of consciousness.
Benign paroxysmal vertigo	Neurological examination and EEG are normal. Abnormal responses are verified by tests of vestibular function.

Differential Diagnosis of Episodic Symptoms

Conditions that must be distinguished from epilepsy are those showing seizures and/or unconsciousness. These conditions and useful findings in differentiating them from epilepsy are listed in Table IV (Pedley et al., 1983). The patient's clinical history and features of the disease are most important for the distinction between epileptic and nonepileptic seizures. EEG monitoring is also useful here. For example, epilepsy is indicated if a flattening of the EEG or an increase in spike-wave activity is detected before the seizure, or if asymmetry is noted postictally with the appearance of slow waves. In some cases, hypoxia is so severe that syncope with convulsive seizure occurs. Polygraphic recording including cardiac monitoring is essential for the diagnosis of seizures due to such syncope. Sleep disorders such as narcolepsy, somnambulism and sleep terrors are distinguished from epilepsy by means of a polysomnogram. Some psychiatric illnesses display pseudoseizures with or without loss of awareness. These seizures are often misdiagnosed as simple or complex partial seizure and occur in patients who have affective disorders and hysteria. Patients who have such psychogenic seizures have a significantly higher incidence of suicidal attempts and psychiatric treatments (Stewart et al., 1982). A combination of epilepsy and psychopathology is seen in some cases. Hospitalization and prolonged simultaneous EEG and video monitoring may be required for the accurate diagnosis of such cases. The aura of panic in patients with complex partial seizures resembles the panic attacks in patients with panic disorder. Clinical history and features will often enable the patient with this disorder to be distinguished from those with epilepsy. Accurate differentiation of these disorders from epilepsy makes it possible to treat patients appropriately and to avoid inappropriate treatment with antiepileptic drugs.

TREATMENT

Drug therapy is the main form of treatment of epilepsy. Surgical operation may be considered for epilepsy which continues in spite of pharmacotherapy.

Drug Treatment of Epilepsy

In the treatment of patients with epilepsy, the most difficult decisions which the physician must make are when to start treatment, which drug to choose, and when to stop therapy.

When to commence antiepileptic treatment. A correct diagnosis of epilepsy is essential for starting antiepileptic drug therapy: a clear distinction between epileptic and nonepileptic attacks is of major importance. Therapeutic trials of antiepileptic drugs in cases of doubt are rarely justified and often make the management of the patient more difficult. An erroneous diagnosis of epilepsy often results in treatment failure.

There have been no adequate studies of the management of a single seizure. A study of prognosis after a first unprovoked seizure showed that the cumulative probability of recurrence was 36% by year 1, 48% by year 3, and 56% by year 5 (Annegers et al., 1986), suggesting that single seizures may not be routinely treated. However, early effective treatment may be important in the prevention of chronic epilepsy. The decision to start therapy, of course, should take into account the needs of the individual patient. If, for instance, the patient operates dangerous machinery or is in a position where a second seizure may permanently disrupt participation in society, antiepileptic medication may be justified.

Treatment of different types of seizure. Monotherapy should be recommended in a newly diagnosed epileptic patient because polytherapy does not suppress seizures more effectively (Lesser et al., 1984; Reynolds, 1987), and because drug toxicity and interactions become increasingly common as the number of drugs administered increases (Collaborative Group for Epidemiology of Epilepsy, 1988). Depending on the response, the dose should be gradually increased until a satisfactory result is achieved or until intolerable adverse effects occur. The choice of the initial and subsequent drugs considered to be appropriate for each seizure type are given in Table V.

In 3 reports on the treatment of tonic-clonic and partial seizures, carbamazepine, phenytoin, and valproate had a higher overall success rate than phenobarbital and primidone (Callaghan et al., 1985; Mattson et al., 1985; Turnbull et al., 1985). Valproate is the most effective for primary generalized tonic-clonic seizures. Partial seizures and secondarily generalized tonic-clonic seizures tend to respond more poorly to treatment than primary generalized tonic-clonic seizures. Carbamazepine may be preferable against these seizures because it has relatively fewer unacceptable adverse effects especially if the starting dose is small and is increased gradually (Trimble and Thompson, 1983; Mattson et al., 1985).

TABLE V. Appropriate Choice of Antiepileptic Drugs

Seizure type	Initial drug	Subsequent drug	
Partial seizure			
Simple partial	CBZ	PHT	
Complex partial	CBZ	PHT	VPA
Secondarily generalized tonic-clonic	CBZ	PHT	VPA
Generalized seizure			
Typical absence	ESM	VPA	
Atypical absence	VPA	CZP	
Myoclonic	VPA	CZP	
Clonic	CBZ	VPA	
Tonic	CBZ	VPA	
Tonic-clonic	CBZ	VPA	PHT

Abbreviations; CBZ: carbamazepine, PHT: phenytoin, VPA: valproate, ESM: ethosuximide, CZP: clonazepam.

Ethosuximide and valproate are the major drugs for the treatment of absence seizures. Almost all patients showed complete control of typical absence seizures (Sato et al., 1982). In patients with absence and tonic-clonic seizures, valproate would be the initial drug of choice since it is effective against tonic-clonic seizures as well. If the patient fails to respond to ethosuximide or valproate alone, combination therapy should be tried (Rowan et al., 1983). EEG is very useful in the assessment of treatment in patients with absence seizures. Clinical response correlates well with the reduction in number of spike-wave paroxysms on the EEG (Penry et al., 1972).

Myoclonic epilepsy is best treated with valproate or benzodiazepines. Most of the myoclonic epilepsies respond well to these drugs, but the Lennox-Gastaut syndrome is refractory.

For infantile spasms, ACTH or corticosteriod therapy has become a conventional form of treatment and there is abundant evidence that immediate improvement in clinical and EEG abnormalities occurs (Hrachovy et al., 1979). However, the long-term prognosis is not influenced by the treatment with these drugs and is usually poor (Pollack et al., 1979).

Monitoring of drug levels. Antiepileptic drugs for which plasma or serum concentration assays are readily available are carbamazepine, clonazepam, ethosuximide, phenobarbital, phenytoin, primidone, and valproate. Assays are particularly useful in checking the drug compliance of the patient. The serum level may provide valuable information when a patient fails to respond as expected to an antiepileptic drug, or when the patient complains of adverse effects of a dose-related nature. There is little doubt that the availability of serum phenytoin assays has greatly improved the clinical use of phenytoin, which is a difficult drug to deal with because of its narrow therapeutic window. Serum level monitoring may also be useful in the treatment with ethosuximide and carbamazepine. Control of absence seizures is related to the serum level of the drug, usually achieved by a level over the usual dose (Sherwin, 1982). Serum level monitoring of other antiepileptic drugs normally of little clinical significance may be of value in situations where renal and hepatic dysfunction or pregnancy alter the protein binding of the drugs.

Adverse effects of antiepileptic drugs. The adverse effects of the chief antiepileptic drugs are well documented. One or more adverse reactions are seen in around 30–40% of patients treated with the drugs for more than 3 months (Collaborative Group for Epidemiology of Epilepsy, 1988). The occurrence and incidence of these adverse effects, of course, varies between drugs. The most common symptom is somnolence, followed by nystagmus, ataxia, gingival hyperplasia, and gastrointestinal symptoms such as nausea, vomiting, and abdominal pain. In the Collaborative Group for Epidemiology of Epilepsy 1988 report, the withdrawal of phenytoin or phenobarbital contributed to the reduction in the number of these adverse effects. They are dose-related and usually reversible. On the other hand, there are idiosyncratic adverse effects whose incidences are usually low but to which careful attention is required. They include, for example, aplastic anemia caused by carbamazepine and phenytoin, fetal hepatic failure by valproate, cerebellar degeneration by phenytoin, and syndrome of inappropriate secretion of antidiuretic hormone by carbamazepine. Clinical findings coupled with serum concentrations provide a reliable basis for the evaluation of drug toxicity.

Teratogenesis. All antiepileptic drugs must be regarded as potentially teratogenic. The incidence of birth defects in children whose mothers took antiepileptic drugs is 2 to 3 times higher than in normal controls (Nakane et al., 1980). The most likely abnormalities are cleft lip and palate, cardiovascular abnormalities, neural tube defects, dysmorphic skeletal abnormalities, and intestinal atresia. High doses of antiepileptic drugs, in particular polytherapy including valproate, are primary factors responsible for the increased incidence of congenital malformation in the offspring of treated epileptic mothers (Kaneko et al., 1988). Of the drugs available, carbamazepine seems to be the safest from this point of view (Janz, 1982). The administration of folate to pregnant women taking phenytoin may prevent such malformations (Zhu and Zhou, 1989).

When to stop antiepileptic treatment. Chadwick (1985) suggests that approximately 20% of children and 40% of adults with epilepsy who have 2–3 year remissions from seizures will have recurrence of seizures on drug withdrawal and most seizures will occur during drug reduction or within the first year after terminating treatment. Despite this result the withdrawal of antiepileptic drugs should be considered in all patients who have been completely seizure-free for at least 2 years with the informed consent of patients. The drugs should be withdrawn slowly to avoid withdrawal seizures. It is impossible to identify groups of patients with particularly high or particularly low risk of relapse on discontinuation of therapy. Some favorable prognostic factors include primarily generalized seizures of short duration without any evidence of neuropsychiatric disorder, seizures of childhood onset, and normal EEG.

Treatment of status epilepticus. Status epilepticus requires immediate medical treatment. In the first place, adequate ventilation and a sure venous line must be secured. Next, 10 mg diazepam or 1 mg clonazepam should be administered intravenously. The administration of these drugs should be limited to 2 or 3 times because the accumulation of diazepam and its metabolites may produce toxicity and because clonazepam causes bronchial hypersecretion. If the status vanishes, a second long-acting antiepileptic drug (phenytoin is recommended) should be administered to avoid recurrence of seizures. Meanwhile, blood examination including serum level of antiepileptic drugs, EEG monitoring, and if necessary other examinations to detect the underlying precipitating causes should be carried out. Metabolic abnormalities should be corrected. If status recurs despite 2 or 3 single-dose administrations, a continuous infusion of chlormethiazole may be tried in spite of a certain risk of respiratory and circulatory depression. Valproate given by nasogastric tube or rectally may be effective. General anesthesia with ventilation may be needed in case of resistance to the above treatment. The administration of antiepileptic drugs must be continued till the restoration of pre-status EEG. The switch to maintenance therapy with oral drugs should be performed with great care by reference to the serum level of the antiepileptic drugs.

Neurosurgery

Temporal lobe epilepsy is the most common seizure for which surgery is used. About 20% of patients with complex partial seizures are candidates for surgery (Mazziotta, 1988). Treatment with antiepileptic drugs must be continued in most patients after the operation. Complete control or a significant decrease in the frequency of seizures is achieved in over 60% of the operated patients (Flanigin et al., 1985). The greatest success is achieved in patients with Ammon's horn sclerosis and those with hamartomas or small tumors.

POSTURAL MYOCLONUS ASSOCIATED WITH LONG-TERM ADMINISTRATION OF NEUROLEPTICS

We have recently noticed rapid, brief myoclonic jerks in the arms and shoulders of schizophrenic patients who had been taking neuroleptics for more than 3 months (Fukuzako et al., 1990). Although the myoclonus developed after neuroleptic withdrawal and was noted in patients with tardive dystonia, no systematic or detailed investigation of such myoclonus has been previously performed.

Sixty patients were investigated for myoclonus and the relationship between postural myoclonus and age, duration of illness, duration of medication, current daily dose, cumulative dose, occurrence of abnormal finger movement, parkinsonism, and tardive dyskinesia were evaluated. Twenty-three patients (38%) showed postural myoclonus when holding the hands forward with the elbow joints flexed at about 90%. Male patients showed a higher incidence of myoclonus than females. The duration of medication is similar for the two groups of patients (with myoclonus 24.1 years vs without myoclonus 21.8 years). Patients with myoclonus had been given significantly higher doses of neuroleptics (chlorpromazine equivalents) than those without myoclonus. There was a significant correlation between the occurrence of myoclonus and abnormal finger movement.

We selected 7 patients with typical myoclonus for detailed study. All 7 patients were given haloperidol and promethazine. However, specific types of neuroleptics causing myoclonus were not identified because other neuroleptics were concurrently administered. Electromyographic recordings in these patients revealed that arrhythmic jerks occurred in the extensor carpi radialis and posterior deltoid muscles and that the jerks on the left and right side were not synchronized (Fig. 10). Clonazepam reduced the frequency of the myoclonic activity.

CONCLUSIONS

In this chapter, we have reviewed mainly the classification, diagnosis, and treatment of epilepsy, taking account of some recent findings. The

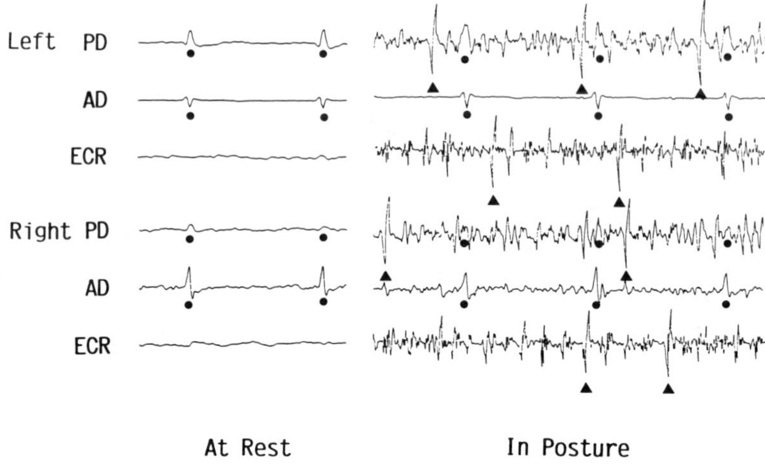

Fig. 10. A postural myoclonus in posterior deltoid (PD) and extensor carpi radialis (ECR) muscle induced by long-term administration of neuroleptics. Positive myoclonic activity (▲) is seen bilaterally in PD and ECR muscle. A solid circle indicates electrocardiographic activity; AD = anterior deltoid. Calibrations are 0.5 sec (x axis) and 100 μV (y axis).

international classifications of epileptic seizures and epileptic syndromes were presented, and the principles on which these classifications are based were described. We also described the signs and symptoms of various types of seizures. Diagnostic evaluations of patients were shown together with figures of EEG, CT, and MRI. Differential diagnoses of episodic symptoms were demonstrated. Practical treatment with antiepileptic drugs and their adverse effects were discussed. Finally, postural myoclonus associated with the long-term administration of neuroleptics was briefly described.

The final goal of treatment is the complete control of seizures. Diagnosis and classification of epilepsy is essential for this purpose; misdiagnosis of nonepileptic seizures as epileptic results in treatment failure. The best treatment must be selected according to the type of epilepsy and the cause of the seizures. New antiepileptic drugs in addition to recent technological developments such as combined video-EEG monitoring, magnetoencepharography (Sato, 1990), CT, MRI, PET, and serum level monitoring of antiepileptic drugs may serve to achieve more effective control of epileptic seizures.

ACKNOWLEDGEMENTS

We are grateful to Drs. H. Tominaga, K. Matsumoto, K. Takeuchi, and F. Nagase (Department of Neuropsychiatry), T. Fukuda (Department of Pharmacology), Drs. K. Sameshima and K. Hiroyama (Department of Pediatrics), and Dr. K. Uchiyama (Department of Neurosurgery) for their cooperation. We should like to thank M. Gore (Kagoshima University) for reading the manuscript and suggesting several improvements.

REFERENCES

Aarts HP, Binnie CD, Smit AM, Wilkins AJ: Selective cognitive impairment during focal and generalized epileptiform EEG activity. Brain 107: 293–308, 1984.

Ajmone-Marsan C, Zivin LS: Factors related to the occurrence of typical paroxysmal abnormalities in the EEG records of epileptic patients. Epilepsia 11: 361–381, 1970.

Annegers JF, Hauser WA, Elveback LR, Kurland LT: The risk of epilepsy following febrile convulsions. Neurology 29: 297–303, 1979.

Annegers JF, Shirts SB, Hauser WA, Kurland LT: Risk of recurrence after an initial unprovoked seizure. Epilepsia 27: 43–50, 1986.

Avrahami E, Cohn DF, Neufeld M, Frishman E, Benmair J, Schreiber R, Korczyn AD: Magnetic resonance imaging (MRI) in patients with complex partial seizures and normal computerized tomography (CT) scan. Clin Neurol Neurosurg 89: 231–235, 1987.

Beaumanoir A: The Lennox-Gastaut syndrome. In Roger J, Dravet C, Bureau M, Dreifuss FE, Wolf P (eds): "Epileptic Syndromes in Infancy, Childhood and Adolescence." London: John Libbey Eurotext, 1985, pp 89–99.

Bergen D, Bleck T, Ramsey R, Clasen R, Ristanovic R, Smith M, Whisler WW: Magnetic resonance imaging as a sensitive and specific predictor of neoplasms removed for intractable epilepsy. Epilepsia 30: 318–321, 1989.

Callaghan N, Kenny RA, O'Neill B, Crowley M, Goggin T: A prospective study between carbamazepine, phenytoin and sodium valproate as monotherapy in previously untreated and recently diagnosed patients with epilepsy. J Neurol Neurosurg Psychiatry 48: 639–644, 1985.

Chadwick D: The discontinuation of antiepileptic therapy. In Pedley TA, Meldrum BS (eds): "Recent Advances in Epilepsy," Vol 2. Edinburgh: Churchill Livingstone, 1985, pp 111–124.

Collaborative Group for Epidemiology of Epilepsy: Adverse reactions to antiepileptic drugs: A follow-up study of 355 patients with chronic antiepileptic drug treatment. Epilepsia 29: 787–793, 1988.

Commission on Classification and Terminology of the International League Against Epilepsy: Proposal for revised clinical and electroencephalographic classification of epileptic seizures. Epilepsia 22: 489–501, 1981.

Commission on Classification and Terminology of the International League Against Epilepsy: Proposal for revised classification of epilepsies and epileptic syndromes. Epilepsia 30: 389–399, 1989.

D'Alessandro R, Pazzaglia P, Tinuper P, Ferrara R, Fabbri R, Lugaresi E: Prognostic and electroclinical features of grand mal epilepsies. Eur Neurol 25: 339–345, 1986.

Daras M, Tuchman AJ, Strobos RJ: Computed tomography in adult-onset epileptic seizures in a city hospital population. Epilepsia 28: 519–522, 1987.

Devinsky O, Kelley K, Porter RJ, Theodore WH: Clinical and electroencephalographic features of simple partial seizures. Neurology 38: 1347–1352, 1988.

Devinsky O, Sato S, Kufta CV, Ito B, Rose DF, Theodore WH, Porter RJ: Electroencephalographic studies of simple partial seizures with subdural electrode recordings. Neurology 39: 527–533, 1989.

Dreifuss FE: Role of intensive monitoring in classification. In Gumnit RJ (ed): "Advances in Neurology," Vol 46. New York: Raven Press, 1986, pp13–26.

Egli M, Mothersill I, O'Kane M, O'Kane F: The axial spasm—The predominant type of drop seizure in patients with secondary generalized epilepsy. Epilepsia 26: 401–415, 1985.

Engel J, Jr, Kuhl DE, Phelps ME, Mazziotta JC: Interictal cerebral glucose metabolism in partial epilepsy and its relation to EEG changes. Ann Neurol 12: 510–517, 1982a.

Engel J, Jr, Brown WJ, Kuhl DE, Phelps ME, Mazziotta JC, Crandall PH: Pathological findings underlying focal temporal lobe hypometabolism in partial epilepsy. Ann Neurol 12: 518–528, 1982b.

Engel J, Jr, Kuhl DE, Phelps ME, Rausch R, Nuwer M: Local cerebral metabolism during partial seizures. Neurology 33: 400–413, 1983.

Engel J, Jr: "Seizures and Epilepsy." Philadelphia: F. A. Davis Company, Ch. 12, 1989, pp 303–339.

Fenton GW: Epilepsy and automatism. Br J Hosp Med 7: 57–64, 1972.

Flanigin H, King D, Gallagher B: Surgical treatment of epilepsy. In Pedley TA, Meldrum BS (eds): "Recent Advances in Epilepsy," Vol 2. Edinburgh: Churchill Livingstone, 1985, pp 297–339.

Fois A, Malandrini F, Mostardini R: Clinical experiences of petit mal. Brain Dev 9: 54–59, 1987.

Fukuzako H, Tominaga H, Izumi K, Koja T, Nomoto M, Hokazono Y, Kamei K, Fujii H, Fukuda T, Matsumoto K: Postural myoclonus associated with long-term administration of neuroleptics in schizophrenic patients. Biol Psychiatry 27: 1116–1126, 1990.

Holden KR, Mellits ED, Freeman JM: Neonatal seizures. I. Correlation of prenatal and perinatal events with outcomes. Pediatrics 70: 165–176, 1982.

Hrachovy RA, Frost JD Jr, Kellaway P, Zion T: A controlled study of prednisone therapy in infantile spasms. Epilepsia 20: 403–407, 1979.

Hughes JR: Two forms of the 6/sec spike and wave complex. Electroencephalogr Clin Neurophysiol 48: 535–550, 1980.

Janz D: On major malformations and minor anomalies in the offspring of parents with epilepsy: Review of the literature. In Janz D, Bossi L, Dam M, Helge H, Richens A, Schmidt D (eds): "Epilepsy, Pregnancy, and the Child." New York: Raven Press, 1982, pp 211–222.

Jeavons PM: West syndrome: Infantile spasms. In Roger J, Dravet C, Bureau M, Dreifuss FE, Wolf P (eds): "Epileptic Syndromes in Infancy, Childhood and Adolescence." London: John Libbey Eurotext, 1985, pp 42–48.

Kaneko S, Otani K, Fukushima Y, Ogawa Y, Nomura Y, Ono T, Nakane Y, Teranishi T, Goto M: Teratogenicity of antiepileptic drugs: Analysis of possible risk factors. Epilepsia 29: 459–467, 1988.

Keränen T, Sillanpää M, Riekkinen PJ: Distribution of seizure types in an epileptic population. Epilepsia 29: 1–7, 1988.

Lerman P: Benign partial epilepsy with centro-temporal spikes. In Roger J, Dravet C, Bureau M, Dreifuss FE, Wolp P (eds): "Epileptic Syndromes in Infancy, Childhood and Adolescence." London: John Libbey Eurotext, 1985, pp 150–158.

Lesser RP, Pippenger CE, Lüders H, Dinner DS: High-dose monotherapy in treatment of intractable seizures. Neurology 34: 707–711, 1984.

Lühdorf K, Jensen LK, Plesner AM: Etiology of seizures in the elderly. Epilepsia 27: 458–463, 1986.

Mattson RH, Cramer JA, Collins JF, Smith DB, Delgado-Escueta AV, Browne TR, Williamson PD, Treiman DM, McNamara JO, McCutchen CB, Homan RW,

Crill WE, Lubozynski MF, Rosenthal NP, Mayersdorf A: Comparison of carbamazepine, phenobarbital, phenytoin, and primidone in partial and secondarily generalized tonic-clonic seizures. N Engl J Med 313: 145–151, 1985.

Mazziotta JC: Practical clinical applications of positron emission tomography in epilepsy. Am J Physiol Imaging 3: 28–29, 1988.

McGahan JP, Dublin AB, Hill RP: The evaluation of seizure disorders by computerized tomography. J Neurosurg 50: 328–332, 1979.

Nakane Y, Okuma T, Takahashi R, Sato Y, Wada T, Sato T, Fukushima Y, Kumashiro H, Ono T, Takahashi T, Aoki Y, Kazamatsuri H, Inami M, Komai S, Seino M, Miyakoshi M, Tanimura T, Hazama H, Kawahara R, Otsuki S, Hosokawa K, Inanaga K, Nakazawa Y, Yamamoto K: Multi-institutional study on the teratogenicity and fetal toxicity of antiepileptic drugs: A report of a collaborative study group in Japan. Epilepsia 21: 663–680, 1980.

Nelson KB, Ellenberg JH: Prognosis in children with febrile seizures. Pediatrics 61: 720–727, 1978.

Pedley TA: Differential diagnosis of episodic symptoms. Epilepsia 24 [Suppl 1]: S31–S44, 1983.

Penry JK, Porter RJ, Dreifuss FE: Ethosuximide: Relation of plasma levels to clinical control. In Woodbury DM, Penry JK, Schmidt RP (eds): "Antiepileptic Drugs." New York: Raven Press, 1972, pp 431–441.

Pollack MA, Zion TE, Kellaway P: Long-term prognosis of patients with infantile spasms following ACTH therapy. Epilepsia 20: 255–260, 1979.

Ramirez-Lassepas M, Cipolle RJ, Morillo LR, Gumnit RJ: Value of computed tomographic scan in the evaluation of adult patients after their first seizure. Ann Neurol 15: 536–543, 1984.

Reynolds EH: Early treatment and prognosis of epilepsy. Epilepsia 28: 97–106, 1987.

Rowan AJ, Meijer JWA, de Beer-Pawlikowski N, van der Geest P, Meinardi H: Valproate-ethosuximide combination therapy for refractory absence seizures. Arch Neurol 40: 797–802, 1983.

Sato S: Magnetoencepharography and epilepsy research. Jpn J Psychiatry Neurol 44: 241–249, 1990.

Sato S, White BG, Penry JK, Dreifuss FE, Sackellares JC, Kupferberg HJ: Valproic acid versus ethosuximide in the treatment on absence seizures. Neurology 32: 157–163, 1982.

Sherwin AL: Ethosuximide: Relation of plasma concentration to seizure control. In Woodbury DM, Penry JK, Pippenger CE (eds): "Antiepileptic Drugs, Second Ed." New York: Raven Press, 1982, pp 637–645.

Sperling MR, Wilson G, Engel J, Jr, Babb TL, Phelps M, Bradley W: Magnetic resonance imaging in intractable partial epilepsy: Correlative studies. Ann Neurol 20: 57–62, 1986.

Stewart RS, Lovitt R, Stewart RM: Are hysterical seizures more than hysteria? A research diagnostic criteria, DSM-III, and psychometric analysis. Am J Psychiatry 139: 926–929, 1982.

Theodore WH, Porter RJ, Penry JK: Complex partial seizures: Clinical characteristics and differential diagnosis. Neurology 33: 1115–1121, 1983.

Theodore WH, Dorwart R, Holmes M, Porter RJ, DiChiro G: Neuroimaging in refractory partial seizures: Comparison of PET, CT, and MRI. Neurology 36: 750–759, 1986.

Tomson T, Svanborg E, Wedlund J-E: Nonconvulsive status epilepticus: High incidence of complex partial status. Epilepsia 27: 276–285, 1986.

Trimble MR, Thompson PJ: Anticonvulsant drugs, cognitive function, and behavior. Epilepsia 24 [Suppl 1]: S55–S63, 1983.

Turnbull DM, Howel D, Rawlins MD, Chadwick DW: Which drug for the adult epileptic patient: Phenytoin or valproate? Br Med J 290: 815–819, 1985.

Wallace SJ: Prevention of recurrent febrile seizures using continuous prophylaxis: Sodium valproate compared with phenobarbital. In Nelson KB, Ellenberg JH (eds): "Febrile Seizures." New York: Raven Press, 1981, pp 135–142.

Young AC, Costanzi JB, Mohr PD, St Clair Forbes W: Is routine computerized axial tomography in epilepsy worthwhile ? Lancet 2: 1446–1447, 1982.

Zhu M-x, Zhou S-s: Reduction of the teratogenic effects of phenytoin by folic acid and a mixture of folic acid, vitamins, and amino acids: A preliminary trial. Epilepsia 30: 246–251, 1989.

Zielinski FF: Epidemiology. In Laidlaw J, Richens A, Oxley J (eds): "A Textbook of Epilepsy, Third Ed." Edinburgh: Churchill Livingstone, 1988, pp 21–48.

2
Neurophysiologic Basis of Convulsive Disorders

WAYNE E. CRILL
*Department of Physiology and Biophysics,
University of Washington, Seattle, WA 98195*

INTRODUCTION

Clinical epilepsy is defined as an increased tendency to have recurrent seizures which are stereotyped behavioral patterns often involving tonic and clonic muscle contractions. Synchronous prolonged electroencephalographic discharges accompany these spells. During the interictal periods, when the epileptic patient is not having seizures, the EEG is useful clinically because it frequently shows intermittent brief high voltage sharp EEG discharges. These interictal spikes are not associated with behavioral symptoms. Neurologists use these clinical symptoms and the EEG patterns to classify epilepsy patients into two general categories. If the ictal or interictal epileptiform EEG discharges appear simultaneously over the entire cerebral cortex the seizure disorder is called *generalized epilepsy.* Many epileptologists infer from this discharge pattern that wide areas of cortex are driven synchronously by diffuse projections from deep cerebral structures, probably in the thalamus or reticular system. A more common clinical form of epilepsy begins with EEG spikes in a focal area that spread for a variable extent through the cerebral cortex. This is called *focal or partial epilepsy* and it is the type of epilepsy most closely related to the seizures studied in experimental laboratories.

Solving the neurophysiologic puzzle of epilepsy requires laboratory models. Numerous agents induce focal seizures in experimental animals. For example, the subpial injection of alumina cream causes focal epilepsy that evolves over several weeks or months with a time course similar to post traumatic epilepsy (Kopeloff et al., 1947; Atkinson and Ward, 1964). Kainic acid is a structural analogue of glutamate and has excitotoxic and convulsive effects (Sloviter, and Damiano, 1981; Westbrook and Lothman, 1983). It is now commonly used as a chronic epilepsy model. Other model systems of epilepsy develop focal epileptic discharges during the course of a single experiment. These acute models are used frequently in the laboratory to study membrane and synaptic properties because the investigator can compare directly normal and epileptic responses in the same tissue or cell. Direct electrical stimulation of brain or the application of drugs such as penicillin, bicuculline, strychnine, or tungstic acid causes the rapid development of focal cortical discharges similar to those present in clinical seizure disorders (Ayala et al., 1970; Ayala et al., 1973; Dichter and Spencer, 1969a; Dichter and Spencer 1969b; Schwindt and Crill, 1981; Schwindt et al., 1984).

The modern era of epilepsy research began with the first description of the electrical events occurring in single neurons during epileptic seizures (Matsumoto and Adjmone Marsan, 1964a; Matsumoto and Adjmone Marsan, 1964b). In the pencillin model, the intermittent spike discharges recorded from the cortical surface have a shape similar to interictal EEG spikes in human focal epilepsy. These experimentally produced interictal discharges are associated with an intracellularly recorded prolonged persistent depolarization often with superimposed action potentials (Fig. 1). This cellular response is called the paroxysmal depolarization shift or PDS. During the seizure itself a depolarization of individual cells persists for several seconds or minutes and often inactivates the neuronal action potential generating mechanism. The transient increase in sodium conductance producing the all-or-nothing spike is inactivated by depolarization. Following the seizure the cell hyperpolarizes and excitability is decreased during the postical period. These cellular events are present in each of the experimental models of epilepsy although there is some variability in the size and shape of the interictal PDS. It is most prominent in penicillin induced seizures.

CELLULAR MECHANISMS DURING EPILEPTIC BEHAVIOR

With the discovery of the PDS and the prolonged seizure depolarization, neurobiologists turned their experimental efforts toward identifying the rules responsible for the PDS and ictal depolarization. An obvious clue to the types of mechanisms is the observation that focal seizure activity spreads into normal brain. There must be intrinsic mechanisms in the brain that can be released by activity in the epileptic focus to cause seizures. Another hint is the similarity of the PDS during interictal responses evoked by a variety of epileptogenic agents suggesting that many different molecular pharmaco-logic actions all ultimately cause the same pathophysiologic response—PDSs and prolonged ictal depolarizations. Even electrical stimulation alone can evoke epileptic discharges. A common epileptogenic factor in these different model epileptic systems is an intense neuronal bursting. The exact sequence of events from high frequency neuronal firing to the spread of synchronized epileptiform bursts is not known but must involve the interplay of the normal synaptic and excitable properties of neurons. In this chapter we will discuss some of these properties and attempt to relate them, so far as is known, to the cascade of excitation that characterizes epileptic behavior.

Membrane Properties of Central Neurons

Since the first description of the cellular responses of individual cells during epileptic activity, neurobiologists have argued whether or not the persistent depolarization during interictal PDS and the ictal event itself were primarily caused by the inward current flowing through ligand-gated channels (synaptic mechanisms) or voltage-dependent channels (membrane mechanisms). Although these discussions still occur it is now clear that epilepsy involves both synaptic and membrane mechanisms.

Action potentials.   The all-or-nothing neuron spikes responsible for the information transfer over distances ranging from microns to meters in the nervous system are generated by the classical Hodgkin-Huxley mechanism (Hodgkin and Huxley, 1952). Neurons have a transient tetrodotoxin (TTX)- sensitive and voltage-dependent sodium conductance that is responsible for brief action potentials. Figure 2 illustrates the sodium spike in cat neocortical cells and the blocking effect of TTX. Action potentials carry information from a neuron's cell body to its presynaptic terminals and the number and frequency of spikes controls the release of neurotransmitter. There is no evidence that the spike generating mechanism per se is altered in epileptic behavior. A potentially significant factor, however, in the generation of seizure discharges is the transduction of graded synaptic signals into a train of action potentials. For example, an enhancement of the input-output properties of a group of neurons could lead to a synchronized avalanche of discharges during the epileptic condition.

One of the remarkable recent discoveries in cellular neurobiology is the large number of different ion channel types in neurons. Every neuron contains other kinds of ion channels in addition to the two Hodgkin and Huxley sodium and potassium channels. For example, there are several types of sodium, calcium and potassium channels, all gated by changes in membrane voltage, in addition to the many kinds of ligand-gated ion channels (Hille, 1984). The structure of these channels and their genetic control is a current area of investigation in many laboratories.

It is ionic current flowing through these channels that generates the electrical responses and the firing patterns of the neuron. The value of membrane potential is determined by the cell's relative permeability for the various ions. Transient depolarizations or hyperpolarizations are caused by changes in selective ionic conductances allowing inward (depolarizing) or outward (hyperpolarizing) ionic currents to flow. One approach in the search for the explanation of interictal and ictal response in neurons is to identify ionic currents that could cause high frequency firing or prolonged depolarizations in neurons. During the early years of cellular epilepsy research it was difficult to identify theoretical ionic mechanisms that might explain ictal and interictal behavior. The only well-studied voltage-dependent ionic channels were the transient sodium and persistent potassium channels so elegantly analyzed by Hodgkin and Huxley (1952), but these mechanisms alone are not sufficient to generate epileptic responses. With the recent discovery of many different ion channel types there is now no shortage of possible mechanisms that could contribute to cellular depolarizations and increased excitability. The challenge is to determine which mechanisms contribute to the synchronous epileptic discharges.

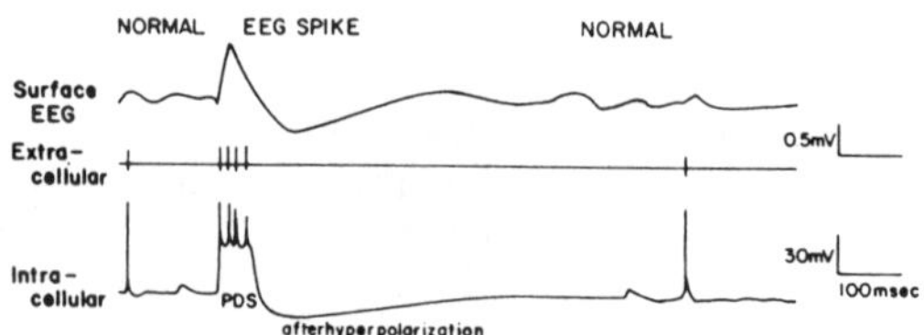
A. INTERICTAL DISCHARGE
NORMAL
EEG SPIKE
NORMAL
Surface EEG
Extra-cellular
0.5mV
Intra-cellular
PDS
afterhyperpolarization
30mV
100msec

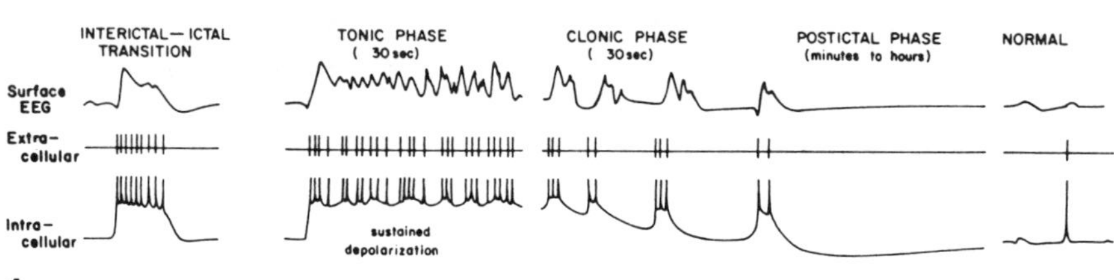
B. SEIZURE DISCHARGE
INTERICTAL—ICTAL TRANSITION
TONIC PHASE (30 sec)
CLONIC PHASE (30 sec)
POSTICTAL PHASE (minutes to hours)
NORMAL
Surface EEG
Extra-cellular
Intra-cellular
sustained depolarization
1

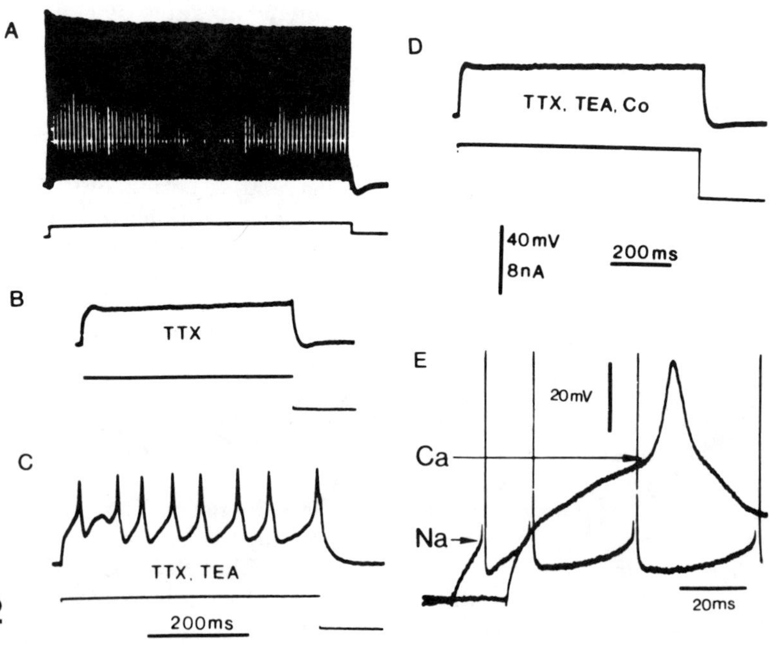
A
B
TTX
C
TTX, TEA
200ms
2
D
TTX, TEA, Co
40mV
8nA
200ms
E
20mV
Ca
Na
20ms

Inward currents. Several different types of depolarizing voltage-dependent currents are activated at membrane potentials negative to spike threshold. Because this is the voltage range for the summation of synaptic potentials the voltage dependent properties of the neuron interact with synaptic inputs and this interaction depends upon the precise membrane potential. Figure 3 illustrates the effect of membrane potential upon the duration of an EPSP in a pyramidal cell from cat neocortex.

Hyperpolarization from resting potential of many different cell types, including neurons, activates a membrane ionic conductance system carrying a depolarizing inward current. Both sodium and potassium ions serve as the charge carrier and because the equilibrium potential for this current is about -20 mV the net current is inward (depolarizing) at resting potential (Mayer and Westbrook, 1983; Spain et al., 1987). This inward rectifier current has been called a number of different names but I_h is the common label. In cardiac cells (called I_f) it is largely responsible for pacemaking of action potentials (Brown and DiFrancesco, 1980). In neurons, the physiologic role of I_h is less well defined. Because the hyperpolarizing inward rectifier, I_h, is largely activated over the voltage range from -60 to -90 mV (Fig. 4) it will tend to keep the cell depolarized from the potassium equilibrium potential which is about -90 mV. In neurons, I_h therefore keeps the membrane potential somewhat depolarized and within striking distance of action potential threshold. I_h is selectively affected by neurotransmitters. Serotonin and norepinephrine increase I_h by stimulating adenylate cyclase (Pape and McCormick; 1989). A membrane voltage-dependent ion channel, I_h is modulated by synaptic mechanisms. I_h also is increased with a rise in $[K^+]_o$. During epileptic seizures $[K^+]_o$ rises to about 10 mM (Somjen, 1984; Somjen and Giacchino, 1985) in the extracellular space which will increase I_h conductance. A larger I_h, caused either by the neurotransmitter modulators or increases in $[K^+]_o$, will move the membrane potential closer to threshold and increase neuron excitability and could change the pattern of neuron firing.

Action potentials are brief because the transient sodium channels characterized by Hodgkin and Huxley (1952) activate and inactivate rapidly. A

Fig. 1. Drawings of electrocorticogram (top trace), extracellular unit recording (middle trace) and intracellular recording (bottom trace) during interictal (**A**) and ictal behavior (**B**). (Reproduced from Lothman and Collins, 1984, with permission of publisher.)

Fig. 2. Sodium and calcium spikes in neocortical neurons in brain slices. **A:** Response of cat neocortical neuron(upper) to constant current stimulation (lower). **B:** Response to a larger constant current stimulation after application of tetrodotoxin (TTX). **C:** Appearance of calcium spikes in presence of TTX to block sodium channels and tetraethylammonium (TEA) ions to block some potassium channels. **D:** Block of calcium spikes with the addition of cobalt (Co). **E:** Relative threshold of sodium and calcium spikes in another neuron. (Reproduced from Stafstrom et al., 1985, with permission of publisher.)

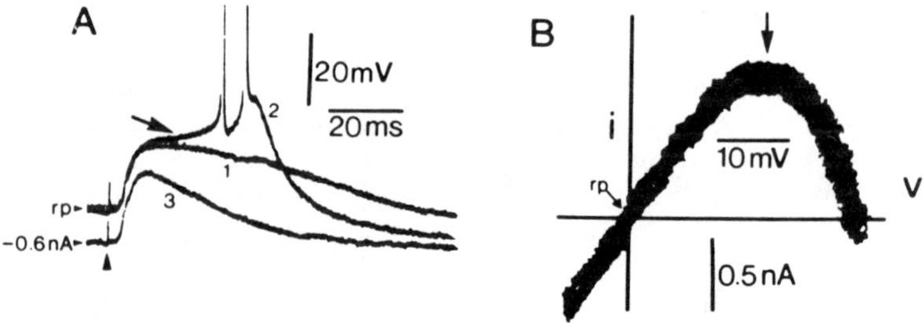

Fig. 3. Altered duration of EPSP as a function of resting potential. **A:** Traces 1 and 2 are superimposed successive responses to evoked EPSPs from normal resting potential. Trace 3 is the response to an identical stimulus with the cell tonically hyperpolarized. **B:** Current-voltage relation in the same neuron shows that downward bend of curve (large arrow) which marks onset of the persistent sodium current and is the value of membrane potential where the EPSP in A is prolonged. (Reproduced from Stafstrom et al., 1985, with permission of publisher.)

surprising twist to the sodium mechanism of neuron excitability was the discovery of low threshold prolonged sodium-mediated depolarizations in cerebellar Purkinje cells. Llinas and Sugimori (1980) attributed this electrogenic response to a noninactivating sodium conductance because it was abolished by tetrodotoxin. Subsequent work in other laboratories has identified similar responses (Connors et al, 1982) and the underlying ionic current (Stafstrom et al., 1982; Stafstrom et al., 1985; French and Gage,

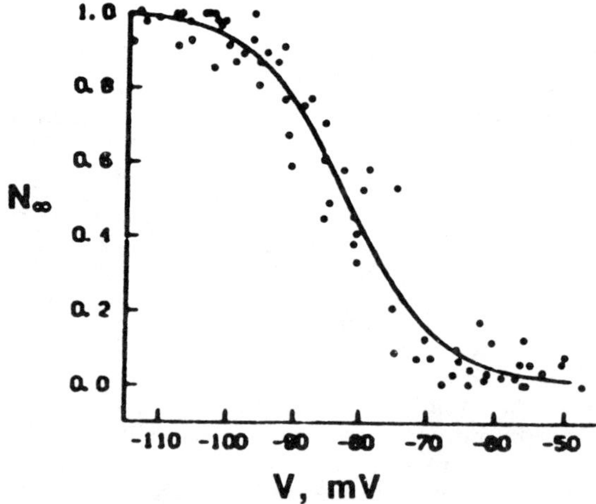

Fig. 4. Relative steady state activation of (I_h) in cat neocortical neurons (N) as a function of membrane potential. (Reproduced from Spain et al., 1987, with permission of publisher.)

1985) in neocortical and hippocampal pyramidal cells. In cat neocortical cells and hippocampal pyramidal cells the persistent sodium current (I_{NaP}) is activated at subthreshold membrane potentials. Figure 5 illustrates the near steady state current- voltage relationship in cat neocortical cells. I_{NaP} is activated rapidly and shows little evidence of inactivation. Because of these kinetic properties, an excitatory postsynaptic potential evoked at potentials near threshold, about 10 mV positive to rest, are prolonged (Fig. 3). Tonic depolarization of neurons toward spike threshold will activate I_{NaP}. None of the experimental epileptogenic agents has a known direct effect upon I_{NaP}, but the relevant observation is that it is present in neurons. As noted above it is the net ionic current that controls neuron polarization. Any decrease in the multitude of repolarizing currents will cause a relative increase in I_{NaP} and alter the input-output relations of neurons.

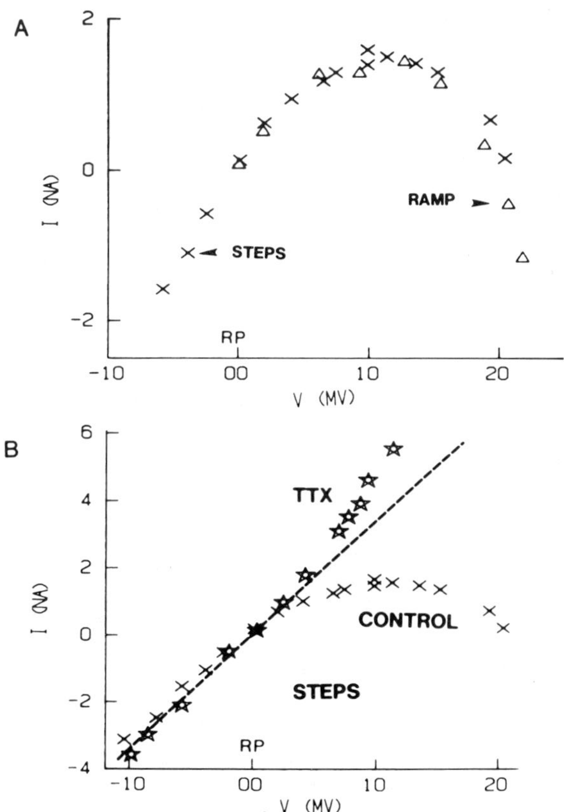

Fig. 5. Voltage dependence of persistent sodium current in cat neocortical neurons. A: Steady state activation of I_{NaP} by voltage-clamp steps and ramps. B: Effect of TTX (10^{-6}M) on I_{NaP}. (Reproduced from Stafstrom et al., 1985, with permission of publisher.)

Calcium currents.   Calcium ions control many regulatory functions in all types of cells that participate in the electrical signalling. The focus here is on the electricity produced by calcium entry through channels in neuronal membranes even though this may only be part of the contribution of calcium mechanisms to the prolonged bursts and synchronization of neurons during epileptic behavior.

All neurons have a family of calcium channels in their cell membranes (Fox et al., 1987). Because intracellular $[Ca^{2+}]_i$ is only about $10^{-7}M$ the opening or activation of these voltage dependent channels allows calcium ions to flow down their electrochemical gradient and depolarize neurons. Three different calcium conductance mechanisms have been identified in neurons. The consequent rise in intracellular $[Ca^{2+}]_i$ has many effects including control of channel gating and transmitter release.

T-type calcium channels are activated at membrane potentials positive to about-70mV and show a relatively rapid inactivation (i.e., close with maintained depolarization). Their opening kinetics is relatively slow and T-channels are not prominent in all neuron types. In the mammalian nervous system the low threshold calcium responses have been characterized most clearly in neurons of the inferior olive nucleus and thalamus (Jahnsen and Llinas, 1984a; Jahnsen and Llinas, 1984b; Llinas and Yarom 1981a; Llinas and Yarom 1981b). In the latter neuron type T-channels contribute to a voltage-dependent regular bursting of the neurons. A depolarizing stimulus applied from a negative potential, where the channel is not inactivated, will transiently open T-channels causing a burst of sodium-dependent action potentials. The T-channels inactivate removing part of the depolarizing current and the neuron repolarizes. Inactivation of the T-channel is removed by the hyperpolarization and the stimulus again evokes a burst of action potentials. The rhythmic firing of thalamic neurons has long been suspected as part of the explanation for generalized epilepsy. Therefore a seminal recent discovery was the demonstration that drugs affecting generalized seizure disorders like ethosuximide selectively decrease calcium current flowing through T-channels (Coulter et al., 1989).

Another type of calcium channel is not activated until a neuron is depolarized to about-20 to -10mV. This high threshold calcium channel (L-channel) has a large unitary conductance (Fox et al., 1987) and is present in every type of neuron examined so far. The L-current is blocked by dihydropyridines and shows very little voltage-dependent inactivation. In most neurons current flowing through L-channels contributes little to the electrogenic response of the cell. Nonetheless, some calcium enters the neuron with each action potential and during high frequency firing intracellular calcium will increase. Some potassium channels are activated by increases in intracellular$[Ca^{2+}]_i$.

Neurotransmitters, like catecholamines in some neuron types can directly decrease calcium currents (Dunlap and Fishbach, 1981). The large and persistent depolarizations occurring during the PDS and the ictal response must activate some of the high threshold calcium conductance allowing calcium to enter the neuron. A measure of calcium entry into neurons during epileptic discharges is the drop in extracellular $[Ca^{2+}]_o$ which occurs during synchronized bursting in the hippocampus (Krnjević et al., 1980).

A less well defined calcium channel type is called the N-channel. They are prominent in dorsal root ganglion neurons and have a unitary conductance between T- and L- values. N-channels are blocked by ω-conotoxin. N-Channels are the major entry for calcium mediated transmitter release in some preparations.

Outward repolarizing potassium currents. As neurobiologists and biophysicists study the ionic channels in neurons (and other cell types), a striking finding is the number of potassium channel types. In the initial axon model of axon excitability (Hodgkin and Huxley, 1952), depolarization caused by an inward sodium current also opened potassium channels. The efflux of potassium ions plus the inactivation of the sodium influx repolarizes the action potential. Repolarization removes the voltage-dependent activation of the potassium conductance over several milliseconds and is reflected in the action potential afterhyperpolarization (AHP). This potassium current is called the delayed rectifier or I_K.

The Hodgkin and Huxley model, which accurately reproduced the experimental responses recorded from axons, does not show the large dynamic repetitive firing range recorded from most neuron cell bodies. Connor and Stevens (1971) provided the first quantitative analysis of an additional potassium conductance mechanism that explained the slow repetitive firing frequencies observed with small stimuli. They discovered a transient potassium conductance mechanism (I_A) that had both voltage-dependent activation and inactivation properties. I_A is activated at subthreshold potentials. A large afterhyperpolarization following the action potential removes inactivation of I_A. As the delayed rectifier decreases following the spike, the neuron slowly depolarizes during the AHP, I_A is activated and the influx of potassium ions slows the firing frequency of the neuron. This is one mechanism for potassium currents to mold the pattern of spike activity in neurons in response to synaptic excitation.

We will not discuss the diversity of these channels except to note that at least six potassium conductance mechanisms have been identified and there is probably an even larger family of ion channels with relatively selective potassium permeability (Stühmer et al., 1989). Several functionally different transient potassium channels like those analyzed by Connor and Stevens (1971) are present in mammalian neurons. At least two other distinct potassium conductance mechanisms are primarily activated by increases in intracellular calcium concentration (calcium mediated potassium channels) and some potassium channels are appreciably modulated by neurotransmitters (Schwindt et al., 1988a; Schwindt et al., 1988b). For example, the m-current is a voltage-dependent potassium current that is reduced by muscarinic agonists (Brown and Adams, 1980). Experimental evidence now points toward different neuron types having qualitatively similar families of potassium channels but the quantitative mixture of types is quite different. It is the potassium channels that shape the output response of neurons to synaptic excitation. They are responsible for the wide variations of the repetitive firing responses of neurons. But what is the role of these many potassium channels in epileptic behavior? With the discovery of multiple inward currents, particularly those that show little evidence for inactivation, and the characterization of the large family of potassium currents, hypothetical "paper physiology" explanations for epileptic-like behavior are easy to formulate. As discussed above, a necessary early step in the development of epilepsy is increased synchronous firing of neurons regardless of the cause. Potassium currents are the ionic mechanism that could suppress this abnormal pattern of firing. They determine the instantaneous firing rate; the adaptation of repetitive firing and the pattern of action potential responses to excitation (Schwindt et al 1988a; Schwindt et al., 1988b). In general, any mechanism that decreases the current flowing through the potassium channels will increase a neurons responsiveness. Decreases in potassium conductance could be a direct action of epileptogenic agents. For example, activation of muscarinic receptors on some neuron types directly decreases potassium currents (the m-current). In many neurons of the cerebral cortex relatively prolonged potassium currents are activated by increases in $[Ca^{2+}]_i$. These calcium activated potassium currents play a major role in controlling the repetitive firing properties of neurons (Schwindt et al., 1988b). Activation of β-adrenergic receptors decreases these currents (Fig. 6). Similarly a reduction in calcium influx will indirectly decrease calcium-mediated potassium currents because intracellular calcium concentration is decreased. Potassium currents can also be reduced without a direct effect upon potassium conductance (Rutecki et

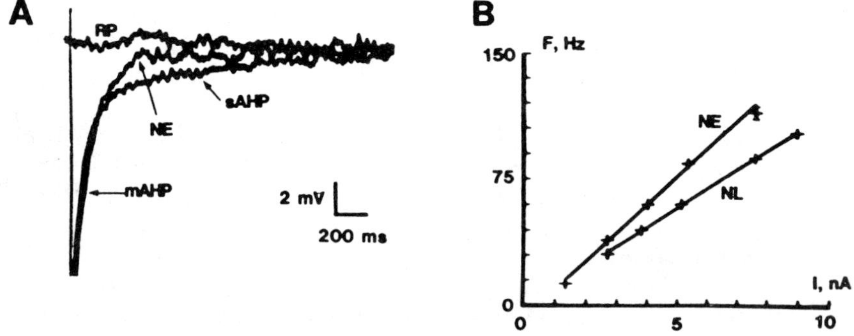

Fig. 6. Effect of norepinephrine (NE) on the afterhyperpolarization (AHP) of cat neocortical neurons. **A:** Superimposed AHPs before (sAHP) and after NE. mAHP is the medium duration afterhyperpolarization. **B:** Firing frequency as function of applied stimulus current in a cat neocortical cell. Norepinephrine (NE) which blocks a portion of the slow AHP increases firing to a tonic current. (Reproduced from Foehring et al., 1989, with permission of publisher.)

al., 1985). It is now well recognized that in the limited extracellular space of brain $[K^+]_o$ increases to 10–12 mM during epileptic activity (Fisher et al., 1976). This change alone moves E_K in a depolarizing direction and current flowing through potassium channels will decrease because the potassium driving potential (membrane potential-E_K) is reduced. Whether or not this change in repolarizing currents is sufficient to allow the normally present inward currents to dominate the responses of the cells is unknown; at a minimum it will lead to a more vigorous response to excitatory input.

Selected Synaptic Properties

Inhibition. This text is devoted to the role of GABA in epilepsy. Therefore the reader is referred to the chapters of interest regarding this topic. Here we will only comment upon the general role of inhibitory synaptic transmission and the pathophysiology of epilepsy. With the discovery of the PDS and the persistent depolarization of neurons during experimental epilepsy, the role of synaptic transmission had great experimental interest. Johnston and Brown (1984) clearly demonstrated that the PDS has properties compatible with a large excitatory synaptic potential. All neuron circuits have extensive excitatory and inhibitory connections. If synaptic inhibition were selectively reduced, excitation could dominate the circuit's responses leading to synchronization and large synaptic depolarizations. Penicillin, one of the common acute epileptogenic agents decreases synaptic inhibition in central neurons (Curtis et al., 1972) and the $GABA_A$ antagonist bicuculline has powerful epileptogenic activity. We still do not know if alterations in the balance of excitation and inhibition are factors in natural human epilepsy. GABA selectively increases chloride conductance and causes a transient inhibitory hyperpolarization because the equilibrium potential for chloride is slightly negative to resting potential in neurons. Heightened inhibitory activity in neurons could increase intracellular chloride concentration and reduce the effectiveness of subsequent inhibition.

Excitatory amino acids. The role of excitatory amino acids in transmission has received intense experimental concern in the last decade because their putative role in learning and memory, cell death, cell development, and epilepsy. The endogenous excitatory transmitters are glutamate and aspatate. At least three excitatory amino acid (EAA) receptors have been identified. They are named on the basis of EAA ligands; N-methyl-D-aspartate (NMDA), kainate and quisqualate. For our discussion we will refer to the last two types as nonNMDA receptors. When EAAs combine with nonNMDA receptors postsynaptic channels are opened with selective permeability to cations. The conductance of these ligand-gated channels shows little voltage dependence and the nonNMDA mediated synaptic depolarization is rapid. The NMDA component of the EAA evoked synaptic potential can be recorded in isolation by blocking the nonNMDA receptor with CNQX (6-cyano-7-nitroquinoxaline-2,3-dione). In contrast, the current flowing through NMDA gated postsynaptic channels have a strong voltage dependence because the channel is blocked at potentials negative to rest by magnesium ions (Nowak et al., 1984). This block is removed when the postsynaptic membrane is depolarized allowing a depolarizing cation current with a large calcium component to flow into the cell. The current through NMDA receptors in most cell types develops slowly. NMDA receptors are blocked by the antagonist 2-aminophosphonovalerate (2-APV). Glycine is a positive allosteric modulator of NMDA channels (Johnson and Ascher, 1987).

The block of the NMDA receptor by physiological extracellular concentrations of magnesium ions gives the current through this ligand-gated channel type voltage-dependent properties. Moderate depolarization will remove the magnesium block of the NMDA channels and allow cations including calcium to enter the cell (Ascher and Nowak, 1987). NMDA application alone causes neuronal depolarizations reminiscent of the PDS (Fig. 7) (Flatman et al., 1983; Flatman et al., 1986). In epilepsy, activation of NMDA receptors could enhance excitatory postsynaptic potentials and produce synaptically generated PDSs. The possible role of NMDA receptors in generating synchronous bursts has been tested by using specific blockers of NMDA receptors like 2-APV and studying the effects of removing the magnesium block in neurons exposed to low extracellular magnesium concentration. Low $[Mg^{2+}]_o$ augments EPSPs by allowing synaptic current to flow through NMDA channels even at resting potential and leads to spontaneous synchronous discharges in brain slices (Mody et al., 1988). These low-magnesium epileptiform responses are blocked by 2-APV. We still do not know all the factors responsible for the apparent selective activation of NMDA receptors by excitatory neurotransmitters, but high frequency stimulation and the associated cell depolarization are important.

CONCLUSIONS

No identified membrane or synaptic property serves as the single cause of epilepsy. In the discussion above we identified many mechanisms that must contribute to recurrent seizures. Each neuron has several membrane channel types that will support prolonged depolarization. These depolariz-

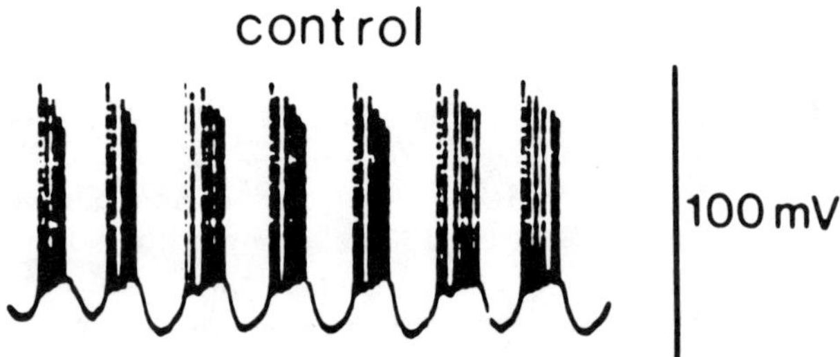

Fig. 7. Rhythmic burst firing induced in a regularly spiking neuron from cat neocortex by the application of NMDA. (Reproduced from Flatman et al., 1983, with permission of publisher.)

ing conductance mechanisms normally do not dominate a neuron's responses because neurons have large repolarizing currents. We know that all neuron circuits have a mixture of excitatory and inhibitory connections. Blocking the inhibitory pathways rapidly leads to epileptic behavior. Many of these membrane and synaptic excitatory mechanisms are enhanced by intense firing in neural circuits. High frequency stimulation raises$[K^+]_o$ because potassium leaves neurons with each action potential. An elevation of $[K^+]_o$ directly increases the inward rectifier and indirectly decreases repolarizing potassium currents by shifting the potassium equilibrium potential in a depolarizing direction. Intense stimulation releases neurotransmitters, such as catecholamines and acetylcholine that decrease calcium mediated potassium currents and increase excitability. High frequency neuronal firing could also decrease the effectiveness of GABA mediated inhibition by shifting the chloride equilibrium potential. The depolarization of neurons during increased firing also will remove the magnesium block of NMDA channels at resting potential. The nervous system, therefore, has many intrinsic neural mechanisms that likely contribute to epileptic bursting and synchronization when they dominate a cell's or a circuit's responses. This at least partially explains why physical means such as electrical stimulation, severe metabolic derangement, or pharmacologic agents can lead to seizures in the normal brain. The most important unanswered question is what is the change in the epileptic focus that periodically releases these "intrinsic epileptic mechanisms" and causes intermittent local spontaneous synchronized firing.

REFERENCES

Ascher P, Nowak L: Electrophysiological studies of NMDA receptors. Trends Neurosci 10:284–288, 1987.

Atkinson JR, Ward AA Jr: Intracellular studies of cortical neurons in chronic epileptogenic foci in the monkey. Exp Neurol 10:285–295, 1964.

Ayala GF, Matsumoto H, Gumnit RJ: Excitability changes and inhibitory mechanisms in neocortical neurons during seizures. J Neurophysiol 33:73–85, 1970.

Ayala GF, Dichter M, Gumnit RJ, Matsumoto H, Spencer WA: Genesis of epileptic interictal spikes. New knowledge of cortical feedback systems suggests a neurophysiological explanation of brief paroxysms. Brain Res 52:1–17, 1973.

Brown DA, Adams PR: Muscarinic suppression of a novel voltage-sensitive K^+ current in vertebrate neurone. Nature 283:673–676, 1980.

Brown H, DiFrancesco D: Voltage-clamp investigations of membrane currents underlying pace-maker activity in rabbit sino-atrial node. J Physiol (Lond.) 308:331–351, 1980.

Connor JA and Stevens CF: Voltage clamp studies of a transient outward membrane current in gastropod neural somata. J Physiol (Lond) 213:21–30, 1971.

Connors BW, Gutnick MJ, Prince DA: Electrophysiologic properties of neocortical neurons in vitro. J Neurophysiol 48:1302–1320, 1982.

Coulter DA, Hugenard JR, Prince DA: Characterization of ethosuximide reduction of low-threshold calcium current in thalmic neurons. Ann Neurol 25:582–593, 1989.

Curtis DR, Game CJA, Johnston GAR, McCulloch RM, MacLachlan RM: Convulsive action of penicillin. Brain Res 43:242–245, 1972.

Dichter M and Spencer WA: Penicillin-induced interictal discharges cat hippocampus. I. Characteristics and topographical features. J Neurophysiol 32:649–662, 1969a.

Dichter M and Spencer WA: Pencillin-induced interictal discharges from cat hippocampus. II. Mechanisms underlying origin and restriction. J Neurophysiol 32:663–687, 1969b.

Dunlap K, Fishbach GD: Neurotransmitters decrease the calcium conductance activated by depolarization of embryonic chick sensory neurones. J Physiol (Lond) 317:519–535, 1981.

Fisher RS, Pedley TA, Moody WJ, Jr, Prince DA: The role of extracellular potassium in hippocampal epilepsy. Arch Neurol 33:76–83, 1976.

Flatman JA, Schwindt PC, Crill WE: The induction and modification of voltage-sensitive responses in cat neocortical neurons by N-methyl-D-aspartate. Brain Res 363:62–77, 1986.

Flatman JA, Schwindt PC, Crill WE, Stafstrom CE: Multiple actions of N-methyl-D-aspartate on cat neocortical neurons in vitro. Brain Res 266:169–173, 1983.

Foehring RC, Schwindt PC, Crill WE: Norepinephrine selectively reduces slow Ca^{2+}- and Na^{+}-mediated K^{+} currents in cat neocortical neurons. J Neurophysiol 61:245–256, 1989.

Fox AP, Nowycky MC and Tsien RW: Kinetic and pharmacological properties distinguishing three types of calcium currents in chick sensory neurones. J Physiol (Lond) 394:149–172, 1987.

French CR, Gage PW: A threshold sodium current in pyramidal cells in rat hippocampus. Neurosci Lett 56:289–293, 1985.

Hille B: "Ionic Channels of Excitable Membranes." Sunderland: Sinauer Assoc., 1984.

Hodgkin AL, Huxley AF: A quantitative description of membrane current and its application to conduction and excitation in nerve. J Physiol (Lond) 117:500–544, 1952.

Jahnsen H, Llinas R: Electrophysiological properties of guinea-pig thalamic neurones: an in vitro study. J Physiol (Lond) 349:205–226, 1984a.

Jahnsen H, Llinas R: Ionic basis for the electroresponsiveness and oscillatory properties of guinea-pig thalamic neurones in vitro. J Physiol (Lond) 349:227–247, 1984b.

Johnson JW and Ascher P: Glycine potentiates the NMDA response in cultured mouse brain neurons. Nature 325:529–531, 1987.

Johnston D, Brown TH: Mechanisms of neuronal burst generation. In Schwartzkroin PA and Wheal H (eds) "Electrophysiology of Epilepsy." London: Academic Press Inc., 1984, pp 277–301.

Kopeloff LM, Barrera SE, Kopeloff N: Recurrent convulsive seizures in animals produced by immunologic and chemical means. Am J Psychiatr 98: 881–902, 1947.

Krnjević K, Morris ME, Reiffenstein RJ: Changes in extracellular Ca^{2+} and K^{+} activity accompanying hippocampal discharges. Can J Physiol Pharmcol 58: 579–583, 1980.

Llinas R, Sugimori M: Electrophysiological properties of in vitro Purkinje cell somata in mammalian cerebellar slices. J Physiol. (Lond) 305:171–195, 1980.

Llinas R, Yarom Y: Electrophysiology of mammalian inferior olivary neurones in vitro. Different types of voltage-dependent ionic conductances. J Physiol (Lond) 315:549–584, 1981a.

Llinas R, Yarom Y: Properties and distribution of ionic conductances generating electroresponsiveness of mammalian inferior olive neurones in vitro. J Physiol (Lond) 315:569–584, 1981b.

Lothman EW, Collins RC: Seizures: In Pearlman AL, Collins RC (eds): "Neurological Pathophysiology." New York: Oxford University Press, 1984, p 234.

Matsumoto H, Adjmone Marsan C: Cortical cellular phenomena in experimental epilepsy: interictal manifistations. Exp Neurol 9:286–304, 1964a.

Matsumoto H, Adjmone Marsan C: Cortical cellular phenomena in experimental epilepsy: ictal manifistations. Exp Neurol 9:305–326, 1964b.

Mayer ML, Westbrook GL: A voltage-clamp analysis of inward (anomalous) rectification in mouse spinal sensory ganglion neurones. J Physiol (Lond) 340:19–45, 1983.

Mody I, Stanton PK, Heinemann U: Activiation of N-methyl-D-aspartate receptors parallels changes in cellular and synaptic properties of dentate gyrus granule cells after kindling. J Neurophysiol 59:1033–1054, 1988.

Nowak L, Bregestovski P, Ascher P, Herbet A, Prochiantz A: Magnesium gates glutamate activated channels in mouse central neurons. Nature 307:462–465, 1984.

Pape HC, McCormick D A: Noradrenaline and serotonin selectively modulate thalamic burst firing by enhancing a hyperpolarized-activated cation current. Nature 31:715–718, 1989.

Rutecki PA, Lebeda FJ, Johnston D: Epileptiform activity induced by changes in extracellular potassium in hippocampus. J Neurophysiol 54:1363–1374, 1985.

Schwindt PC, Crill WE: Voltage-clamp study of cat spinal motoneurons during strychnine-induced seizures. Brain Res 204:226–230, 1981.

Schwindt PC, Spain W, Crill WE: Epileptogenic action of tungstic acid gel on cat lumber motoneurons. Brain Res 291:140–144, 1984.

Schwindt PC, Spain WJ, Foehring RC, Stafstrom CE, Chubb, MC, Crill WE: Multiple potassium conductances and their functions in neurons from cat sensorimotor cortex in vivo. J Neurophysiol 59:424–449, 1988a.

Schwindt PC, Spain WJ, Foehring RC, Crill WE: Slow conductances in neurons from cat sensorimotor cortex in vitro and their role in slow excitability changes. J Neurophysiol 59:450–467, 1988b.

Sloviter RS, Damiano BP: On the relationship between kainic acid-induced epileptiform activity and hippocampal neuronal damage. Neuropharmacology 11:1003–1011, 1981.

Somjen GG: Interstitial ion concentration and the role of neuroglia in seizures. In Schwarzkroin PA and Wheal H (eds): "Electrophysiology of Epilepsy" London: Academic Press, pp 303–341.

Somjen GG, Giacchino JL: Potassium and calcium concentrations in interstitial fluid of hippocampal formation during paroxysmal responses. J Neurophysiol 53:1098–1108, 1985.

Spain WJ, Schwindt PC, Crill WE: Anomalous rectification in neurons from cat sensorimotor cortex in vitro. J Neurophysiol 57:1555–1576, 1987.

Stafstrom CE, Schwindt PC, Crill WE: Negative slope conductance due to a persistent subthreshold sodium current in cat neocortical neurons in vitro. Brain Res 236:221–226, 1982.

Stafstrom CE, Schwindt C, Chubb MC, Crill WE: Properties of persistent sodium conductance and calcium conductance of layer V neurons from cat sensorimotor cortex in vitro, J Neurophysiol 53:153–170, 1985.

Stühmer W, Ruppersberg JP, Schroter, KH, Sakmann B, Stocker, M Giese KP, Perschke A, Baumann A, Pinge O: Molecular basis of functional diversity of voltage-gated potassium channels in mammalian brain. EMBO J 11: 3235–3244, 1989.
Westbrook GL, Lothman EW: Cellular and synaptic basis of kainic acid-induced hippocampal epileptiform activity. Brain Res 273:97–109, 1983.

GABA Mechanisms in Epilepsy, pages 47–87
© 1991 Wiley-Liss, Inc.

3
Significance of GABA in Brain Function

KRESIMIR KRNJEVIĆ
*Anaesthesia Research Department, McGill University,
Montréal, Quebec, Canada H3G 1Y6*

INTRODUCTION

What Is GABA?

γ-Aminobutyric acid (GABA) is a small monocarboxylic acid (~ 100 dalton). It is readily obtained by a α-decarboxylation of glutamic acid. It differs from most biologically-occurring amino acids in having its amino group in the omega position, and not being a constituent of proteins. Why is GABA important? Because so much evidence has now accumulated pointing to GABA as the principal transmitter released by countless inhibitory cells, in all parts of the vertebrate brain.

Brief History

Having no known function, not much attention was paid to the discovery—40 years ago—that GABA is present in large amounts in the brain (Roberts and Frankel, 1950; Awapara et al., 1950). Its possible significance became much more evident several years later when Bazemore et al. (1957) identified GABA as the endogenous substance (Factor I) responsible for the inhibitory action of brain extracts on crayfish neurons. Although it had a powerful inhibitory action on mammalian spinal neurons (Curtis et al., 1959), this was thought to differ significantly from synaptic inhibition. In the cortex, however, GABA's action and IPSP's proved to be practically identical (Krnjević and Schwartz, 1967), and GABA was therefore proposed as the natural inhibitory transmitter. Subsequent advances that helped to establish the widespread nature of GABAergic inhibition included the discovery that bicuculline is a specific antagonist of GABA receptors (Curtis et al., 1971); the isolation of glutamic acid decarboxylase from brain and its use to generate antibodies capable of identifying GABA-producing neurons (McLaughlin et al., 1974); and the recognition that there are separate $GABA_A$ and $GABA_B$ receptors with distinct pharmacology and mechanisms of action (Bowery et al., 1980). A further important finding was that the activation of $GABA_A$ receptors is much enhanced by some widely-used anesthetics (Nicoll, 1972) or benzodiazepine tranquilizers (Choi et al., 1977): this enormously widened the range of GABA studies which now encompasses research on a great variety of neurological (especially convulsant) and psychiatric disorders (McGeer et al., 1987).

GABAERGIC CELLS AND PATHWAYS

A systematic survey of all putative GABAergic cells and synapses would require several volumes. A brief survey of some specific examples will give an idea of some of the modes of connection and operation.

Central Sensory Pathways

At every level GABAergic cells tend to limit, or sharpen, the sensory signal (spatially and temporally) and to enhance its specificity. In the somatic sensory pathway this is evident from the first (spinal) relay, where presynaptic inhibition operates (Eccles, 1964), to the cortex (Dykes et al., 1984). This is no less true of the visual system: at every level from the retina (Massey and Redburn, 1987), through the geniculate to the cortex (Sillito, 1984), GABAergic mechanisms determine not only the amount of signal but the kind of information that it conveys.

In the thalamus, an additional more global type of control is exerted by the GABAergic cells of the reticular nucleus (perigeniculate nucleus for the visual division). There is strong evidence (Mulle et al., 1986; Steriade and Llinas, 1988) that synchronized rhythmic activity of these cells is responsible for the overall synchronization of thalamic activity which manifests itself both in the thalamus and the cortex as "spindles" (or α- rhythm). Though its significance is still not fully evident, this appears to be a mechanism of global disconnection of the thalamo-cortical system from afferent (sensory) inputs.

Motor Systems

Cerebellum. The realization that cerebellar activity is shaped by GABAergic neurons and that its output is purely inhibitory brought about a major conceptual revolution (Ito, 1984). The original discovery (Ito and Yoshida, 1966) that the principal output neurons, the Purkinje cells, generate only monosynaptic IPSPs in the deep cerebellar nuclei and the medulla proved that the Purkinje cells are solely inhibitory, and that inhibitory cells are not necessarily short-axoned, locally-acting neurons. In addition to providing an inhibitory output GABAergic cells are involved in sharpening or in other ways modifying information transfer at every step: the Golgi cells modulate the input received and transmitted by granule cells; basket and stellate cells sharpen the focus of parallel fibre action on Purkinje cells; and the Purkinje cells' own activity is moderated by negative feedback via inhibitory recurrent collaterals (Ito, 1984).

Basal ganglia. Though much more opaque than the cerebellum in its detailed morphological organization, this region presents an even more remarkable arrangement of successive groups of inhibitory neurons

(Yoshida and Obata, 1978; McGeer et al., 1987). Apart from the major input from the cortex, which is excitatory and almost certainly glutamatergic—most of the further information transfer appears to be mediated by GABAergic cells: from the caudate/putamen to the pallidum; then to the reticular portion of the subtantia nigra; and finally to the brainstem. These large nuclei are thus composed largely of GABAergic neurons. Such a chain of inhibitory neurons could operate only if a sufficient level of excitation was somehow maintained, either by intrinsic (pacemaker) activity, or more likely by excitatory agents released by other cells (such as ACh, substance P, etc.).

The substantia nigra is of particular interest in the present context because it appears to play a central role in the development or propagation of generalized seizures (McNamara et al., 1984; Gale, 1985). The precise mechanisms involved are not fully known, but there is good evidence that GABAergic inhibition of nigral cells is an important moderating influence, which tends to reduce susceptibility to generalized seizures initiated in other brain regions.

GABAERGIC INHIBITION

The present article cannot attempt to give an overview of all the relevant literature. Rather it surveys the main features of GABA's role in brain function with special emphasis on some of the more recent findings. For some previous reviews of GABAergic transmission and its pharmacology, several articles can be consulted (Krnjević 1974; 1976; Curtis and Johnston, 1974; Nistri and Constanti, 1979; Olsen 1982; Enna and Gallagher, 1983; Olsen et al., 1984; Ribak, 1987; McGeer et al., 1987; Dingledine et al., 1986; DeFeudis, 1989).

Basic Mechanisms

Inhibitory synapses can operate in two general ways: either by reducing excitatory transmitter release, through a *presynaptic* action, or by reducing the excitability of *post-synaptic* cells (theoretically, some other kinds of actions are possible, such as a specific *antagonism* of excitatory transmitters; but so far they have not proved to be of major importance). GABA indeed has both pre-and post-synaptic inhibitory effects. Both types can be mediated by $GABA_A$ or $GABA_B$ receptors.

The post-synaptic mechanisms are well-understood and of a relatively simple kind; by increasing the membrane conductance for an ion having an equilibrium potential near, or more negative than, the resting membrane potential (V_m), they oppose depolarization and thus prevent cell firing. The important difference is that $GABA_A$ inhibition acts by generating a high Cl^- conductance (G_{cl}) and $GABA_B$ inhibition by raising K^+ conductance (G_K).

The presynaptic mechanisms are not nearly so certain because intracellular recording from most nerve terminals is not practicable- proposed mechanisms are usually based on extrapolation of findings on the parent cell bodies (e.g., dorsal root ganglion cells). Presynaptic $GABA_A$ actions also appear to be mediated by G_{cl} increase. But the presynaptic $GABA_B$ actions may be mediated by a direct depression of terminal Ca current, perhaps reinforced by increased G_K.

$GABA_A$ and $GABA_B$ receptors differ greatly not only in their pharmacological characteristics (agonists, antagonists), but also in a very basic respect: the $GABA_A$ type *directly* activates Cl^- channels whereas the activation of K channels by $GABA_B$ actions is GTP-dependent, and therefore mediated indirectly, via a G-protein (cf. Rodbell, 1980; Cockroft and Stutchfield, 1988).

$GABA_A$ Receptor-Mediated Inhibition

G_{cl} enhancement. That IPSPs are highly sensitive to Cl^- leakage or injection (from recording micropipettes) was evident from the first intracellular studies in the mammalian spinal cord (Coombs et al., 1955), and confirmed in numerous subsequent investigations in the brain (Kandel et al., 1961; Eccles, 1964; Krnjević, 1976). Indeed, the fact that Cl^- injections produced a similar depolarizing shift in reversal potential both for IPSPs and for the effect of GABA was a major step in identifying GABA as the natural inhibitory transmitter in the cerebral cortex (Krnjević and Schwartz, 1967; Dreifuss et al., 1969). Similar data were later obtained in the hippocampus in slices (Andersen et al., 1980) and in situ (Ben-Ari et al., 1981a), as illustrated in Figure 1.

However, further attempts to assess more precisely the permeability of these inhibitory channels by intracellular injections of various ions led to some controversy. From their observations on IPSPs in spinal motoneurons in cats, Coombs et al. (1955) and later Araki et al. (1961) and Eccles (1964) concluded that the inhibitory channels are permeable to small cations (such as K^+), as well as all anions up to the size of formate; an upper limit of 3.4Å was proposed for the pore diameter (calculated from the ionic limiting conductivity and therefore the *apparent* size of the hydrated ions). Although these IPSPs are mediated mainly by glycine, the underlying conductance changes are practically identical to those produced in the same cells by GABA (Hamill et al., 1983; Bormann et al., 1987). However, in cortical cells, (also in cats) inhibitory channels seemed to be purely anionic and significantly larger (5–8Å), with evidence of a small but significant permeability to HCO_3^- and acetate, and even some substantially larger anions (Kelly et al., 1969).

As a result of extensive single channel studies by the patch clamp technique on cultured spinal cells (Hamill et al., 1983; Bormann et al., 1987; Bormann, 1988) it is now clear that the $GABA_A$ (or glycine)-activated channels are indeed purely anionic and that they have an effective

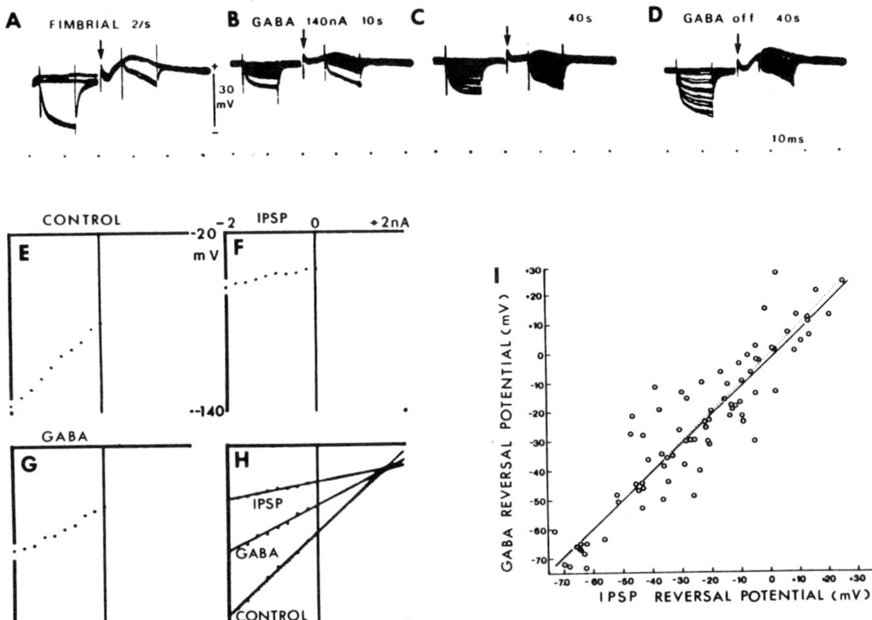

Fig. 1. Similar inhibitory action of GABA and IPSPs on pyramidal cells of rat hippocampus in situ. **A-D:** superimposed oscilloscope traces show IPSP and associated resistance drop evoked by fimbrial stimulation (at arrows); throughout, incrementing series of 10 ms hyperpolarizing current pulses were injected into cell both before and after fimbrial stimulus to measure changes in resistance. **A:** Control traces; note IPSP is in depolarizing direction because recording electrode contained 3M KCl; **B:** during GABA iontophoretic application, there is a depolarizing (upward) shift of baseline and marked reduction of resistance; **C:** after 40 s of GABA application, its action is diminished ("fading") and IPSP conductance increase is also much smaller. **D:** Shows recovery of resting resistance and IPSP 40 s after end of GABA release. **E-H:** Voltage current plots from similar experiment show **E.** resting values **F,G:** depolarization and marked drop in resistance at peak of IPSP and GABA action respectively and **H**, superimposed data from **E-G**, with best fitting regression lines: similar intersections of extrapolated lines for GABA and IPSPs indicate similar reversal potential. **I.** graph summarizing comparable data obtained from 27 hippocampal neurons shows excellent agreement between corresponding reversal potentials for IPSPs and for GABA action: solid line is line-of-best-fit; dotted line that of perfect agreement. (Reproduced from Figs. 11, 13, and 14 in Ben-Ari et al., 1981a with permission of publisher.)

diameter of 5-6Å—in relatively good agreement with the earlier studies on cortical cells in situ. With neither glycine nor GABA was there evidence of any significant K conductance. Only minimal differences could be detected between glycine and GABA-activated channels; for example both types showed an anomalously high permeability to SCN^-, as well as a significant permeability to acetate and HCO_3^- (Fig. 2). The reverse order of relative *conductances* (as opposed to *permeabilities*) suggested very similar pore structures, with relatively strong binding of SCN^- *within* the pores. Another point of interest was the consistent appearance of subconductance states of both GABA-and glycine-activated Cl^- channels; and there was clear evidence of outward rectification, indicating some voltage-dependence of channel gating (Fig. 3).

An important consequence of the substantial HCO_3^- permeability of GABA/glycine-activated channels (estimates of P_{HCO_3}/P_{Cl} have ranged from 0.2 to 0.4) (Kelly et al., 1969; Bormann et al., 1987; Kaila et al., 1989) is that inhibition may generate quite large transmembrane movements of HCO_3^- (Fig. 4). The resulting changes in pH$_i$ (Kaila et al., 1989) could have significant functional connotations. The converse is also true: V_{HCO_3} is

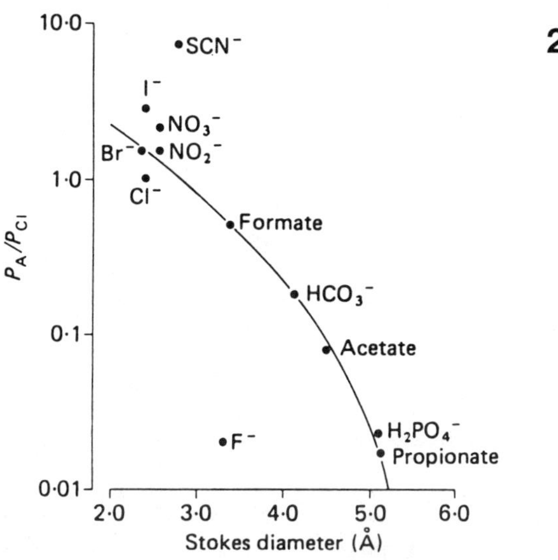

Fig. 2. Relation between permeability ratio and apparent ionic diameter for various anions permeant through GABA-activated channels of cultured mouse spinal cord neurons. Relative permeability ratios of various anions were calculated (with respect to Cl^-) from bi-ionic reversal potentials. Ionic diameters were calculated from limiting conductances of the ions. The data were scaled according to the permeability ratio measured with formate. The diameter of the pore in this model was chosen as 5–6 Å to give the fit shown by the continuous line (Reproduced from Fig. 7 in Bormann et al., 1987, with permission of publisher.)

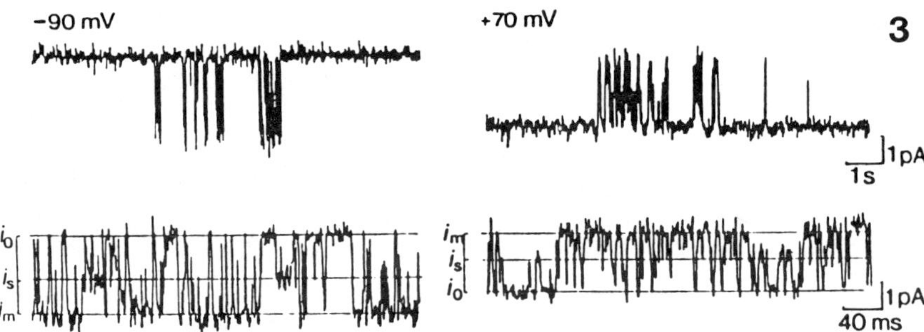

Fig. 3. Bursts of GABA receptor channel currents in cultured spinal cord neurons (from mouse embryos). Outside-out patch records from different patches activated by desensitizing concentration of GABA 20 μM. Current bursts were recorded at -90 mV and +70 mV. Internal K^+ was replaced by $Tris^+$ to allow study of Cl^- channels in isolation from K^+ channels that are activated by depolarization. The initial increase in current caused by GABA gradually diminished as channels dropped out due to desensitization. Then bursts of current lasting 0.2–2 s, which occurred in clusters of 2–10 bursts, could be recorded (upper traces). Bursts, which are separated by long silent periods of up to 20 s, probably reflect the open-closed transitions of individual channels as they transiently leave the desensitized state. Examination of events within burst at higher resolution (lower traces) indicated brief openings of 1–10 ms that were interrupted by brief (< 2 ms) closing gaps. Two distinct current levels were evident while the channel was open: a main current level (i_m) of 2.9 pA, and a sublevel (i_s) of 1.7 pA. At +70mV similar clusters of outward current bursts were also recorded with a main current level of 2.2 pA and a sublevel of 1.2 pA (day 37 neuron). (Reproduced from Fig. 2 in Hamill et al., 1983, with permission of the publisher.)

typically much more positive than V_{cl}; therefore the G_{HCO_3} component of the IPSP could explain some depolarizing GABAergic actions, without the need to postulate inward Cl^- transport. Since $[HCO_3^-]_i$ is wholly determined by pH_i (Stewart 1981), V_{IPSP}, and therefore the potency of inhibitory synapses, could be markedly influenced by metabolic changes that alter pH_i. As pointed out by Kaila et al. (1989), postsynaptic alkalosis would reduce, and acidosis enhance, the efficacy of IPSPs. The positive shift in V_{IPSP} produced in the cortex by NH_4^+ (Lux et al., 1970; Raabe and Gumnit, 1975) is fully in keeping with this idea. [The lack of comparable effect of NH_4^+ in the hippocampus (Allen et al., 1977) may be due to stronger buffering power of hippocampal neurons].

A significant contribution of HCO_3^- to IPSPs is of more than purely academic interest. Though generally believed to have a hyperpolarizing effect— because V_{cl} in the brain is usually more negative than V_m—in fact GABAergic IPSPs can be predominantly or even purely depolarizing potentials. Of course, when V_{GABA} is more negative than V_m large applications of GABA, or high frequency activation of inhibitory cells, will generate a large influx of Cl^- and therefore hyperpolarizing IPSPs rapidly diminish or even disappear. But in certain neurons, such as hippocampal granule cells, IPSPs

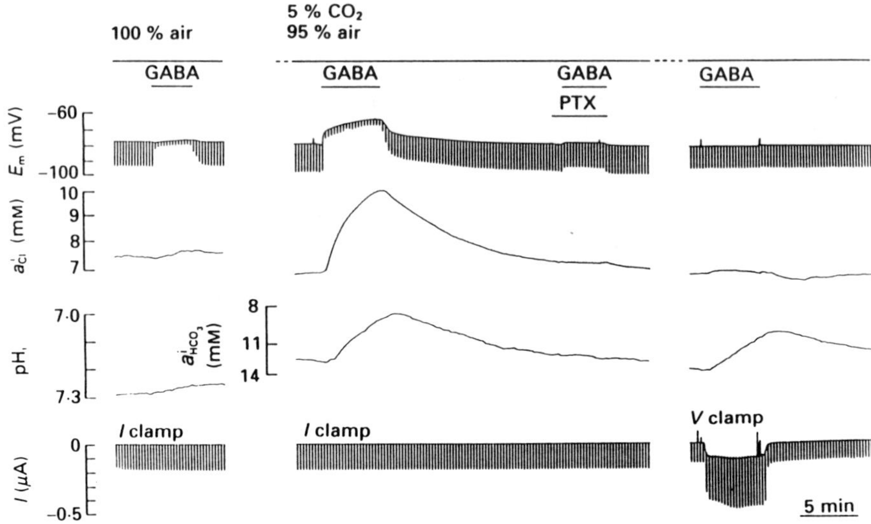

Fig. 4. Effects of GABA on membrane potential (E_m), current (I), intracellular Cl^- and HCO_3 activities (a^i_{cl} and Cl^i_{HCO3}) and pH_i of crayfish muscle fibre. Left and middle: in the presence of HCO_3^- (5% CO_2), the increase in conductance produced by 0.2 mM GABA is paralleled by an increase in a^i_{cl} and by a fall in $a^i_{HCO_3}$ (fall in pH_i). Both effects are blocked by picrotoxin (0.1 mM). Right: the GABA-induced increase in a^i_{cl} (but not the fall in $a^i_{HCO_3}$) is inhibited by clamping E_m at its resting value; evidently, at this potential the GABA-induced current is mediated mostly by HCO_3^- efflux. The pH_i scale applies throughout the experiment. Calibration of $a^i_{HCO_3}$ (in the presence of CO_2) is based on pH_i reading. The breaks between the recordings were 90 min. and 10 min. I clamp = current clamp; V clamp = voltage clamp. (Reproduced from Fig. 4 in Kaila et al., 1989, with permission of the publisher.)

are consistently in the depolarizing direction (V_{IPSP} = -62 mV) (Misgeld et al., 1986); whereas in the same preparations, CA3 pyramidal cell (somatic) responses are mainly hyperpolarizing and V_{IPSP} is significantly more negative -70 mV (Misgeld et al., 1986). Moreover, dendritic responses tend to be either mainly depolarizing (Alger and Nicoll, 1982; Andersen et al., 1980; Avoli and Perreault, 1987) or biphasic (Misgeld et al., 1986). A different type of $GABA_A$ receptor has been proposed as the explanation for dendritic depolarizing responses (Alger and Nicoll 1982). An alternative possibility is that V_{cl} varies in different cells (or even in different regions of the same cell), depending on the relative importance of outward and inward Cl transport (Misgeld et al., 1986). But a perhaps even simpler explanation might be a somewhat higher G_{HCO3} or pH_i in granule cells and pyramidal dendrites. Similarly, a marked drop in relative G_{HCO3} or pH_i could account for the striking developmental change seen in hippocampal pyramidal cells at the end of the first post-natal week (in brain slices from rats), when GABA ceases to evoke large depolarizations (Ben-Ari et al., 1989).

Whether [Cl]$_i$ as such is really *controlled* is far from evident. Perhaps because intracellular mechanisms (enzymes and second messenger systems, for example) are not particularly sensitive to Cl⁻, most cells may not have acquired a very effective primary pump for Cl⁻. The prevailing [Cl]$_i$ seems to be the resultant of passive Cl⁻ influx and efflux through Cl channels (in nerve cells seldom open except during synaptic inhibition) and such active mechanisms as K⁺-Cl⁻-or Na⁺-K⁺-Cl⁻ cotransport and Cl⁻: HCO⁻₃ exchange (Fig. 5), as well as perhaps a primary Cl⁻, ATP- driven transport (Shiroya et al., 1989)- for an extensive discussion of Cl⁻ transport in nerve cells see Alvarez-Leefmans (1990).

Pharmacological properties. There have been numerous pharmacological studies of GABA$_A$ receptors. A remarkable variety of drugs have proved to be potent agonists , antagonists, and, more surprisingly, modulators (Olsen, 1982; Enna and Gallagher 1983; McGeer et al., 1987; DeFeudis, 1989).

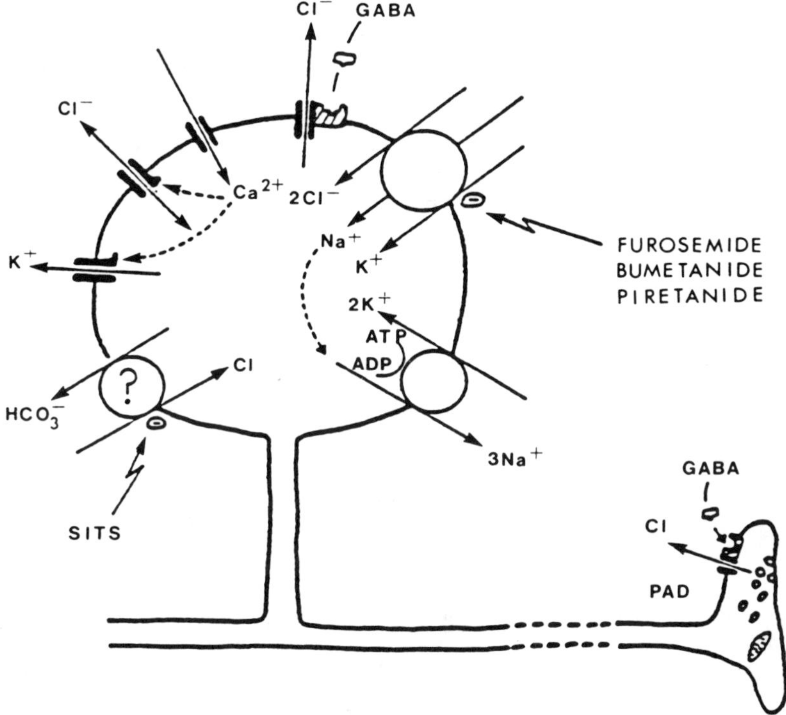

Fig. 5. Diagram illustrating the postulated mechanisms of Cl⁻ movement in dorsal root ganglion cells: GABA- or Ca²⁺-evoked Cl⁻ efflux; coupled Na⁺, K⁺ and Cl⁻ inward transport (blocked by furosemide, for example) and Cl⁻—HCO₃⁻ exchange (blocked by SITS). (Reproduced from Fig. 13 in Alvarez-Leefmans, 1990, with permission of the publisher.)

Agonists and antagonists.   The important point to note is that several agonists (e.g., muscimol, THIP, isoguvacine) are much more specific than GABA itself, for practical purposes having no effect on GABA$_B$ receptors. On this account, muscimol, which has a particularly high affinity for GABA$_A$ receptors and readily passes through the blood brain barrier, has been widely used for binding studies of the distribution of GABA$_A$ receptors, throughout the central and peripheral nervous system (DeFeudis, 1982).

Further evidence of the specific nature of GABA$_A$ receptors is the existence of a number of selective antagonists. These include such well-known competitive antagonists as bicuculline and penicillin (Curtis and Johnston, 1974), as well as a variety of non-competitive blockers, of which picrotoxin has longest been in use. In fact, many convulsants—including metrazol (Macdonald and Barker, 1977) and tubocurarine (Lebeda et al., 1982)—have proved to be GABA$_A$ antagonists. The most obvious exception is strychnine (and some related drugs) which act selectively at glycine receptors (Curtis and Johnston, 1974), and therefore are particularly effective in the spinal cord and lower brain stem level.

Modulators.   GABA$_A$ receptors are of the greatest importance for brain function because GABAergic synapses are so ubiquitous and constantly active. But interest in their function was enormously enhanced by the discovery that they can be "modulated" by a wide variety of psychoactive drugs, notably barbiturates (Nicoll, 1972) and benzodiazepines (Choi et al., 1977).

Such modulation—or more precisely potentiation—of the inhibitory action of GABA was highly significant, in two ways. Firstly, because it cast quite new light on the basic properties of the GABA receptors, which came to be seen as complex macromolecules that include not only the GABA$_A$ receptor site and the Cl$^-$ channel, but in addition have separate binding sites for barbiturates and benzodiazepines (Fig. 6).

Secondly, it opened up a new world of clinical GABAergic pharmacology. It can be taken as a truism that simple agonists or antagonists of the main synaptic transmitters are likely to have only very limited therapeutic usefulness, because their effects are too powerful and undiscriminating. By contrast, modulatory drugs, which only potentiate or diminish transmitter actions, are far more likely to be usable in a controlled and reasonably safe manner.

Of course, "allosteric" regulatory sites on other macromolecules (such as enzymes) have been known for a long time (Monod et al., 1965). At such sites, which are distinct from the catalytic site where substrates bind, other agents that bind cause a change in the conformation of the macromolecule such that the affinity for substrates is greatly enhanced or the speed of reaction (V_{max}) much accelerated. Through comparable conformational changes, various modulators of the GABA$_A$ receptors can alter the affinity for GABA and its ability to open the Cl$^-$ Channel. In precisely what manner these changes in receptor shape and/or charge distribution take place will become increasingly clear as details of the molecular structure and properties become more and more evident (see page 60).

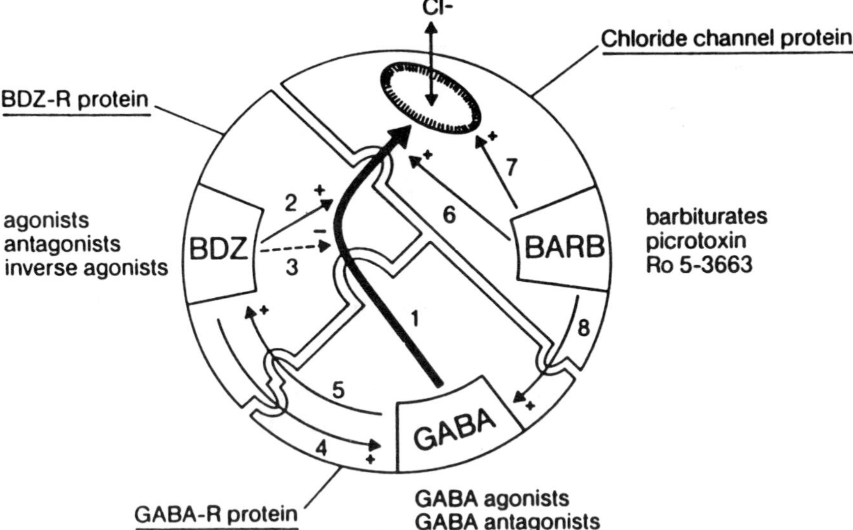

Fig. 6. Model of the GABA$_A$ receptor complex, incorporating electrophysiological and biochemical evidence. Binding sites for GABA, benzodiazepines (BDZ) and barbiturates (BARB) are allosterically coupled within this three-receptor protein complex. *Electrophysiology*: In the process of GABA receptor activation (**1**) leading to increased Cl$^-$ conductance, BDZ antagonists bind to BDZ binding sites and antagonize the effects of BDZ agonists (**2**) and BDZ inverse agonists (**3**). BARB increase the GABA-induced gating close to Cl$^-$ channels (**6**) and, in higher concentrations, directly open the channel (**7**). In contrast, picrotoxin and the convulsant Ro 5–3663 block the channel function. *Biochemistry*: BDZ agonists enhance the GABA receptor affinity (**4**) whereas GABA agonists increase the binding of BDZ agonists (**5**), BARB may also enhance the GABA receptor affinity in a picrotoxin-sensitive manner (**8**). (Reproduced from Fig. 4 in Polc et al., 1982, with permission of the publisher.)

Modulators: Barbiturate site. When they bind to the GABA receptor, a variety of barbiturates either enhance the action of GABA (Figs. 7 and 8), or, in high concentration, open the Cl$^-$ channel (Macdonald; et al., 1986; Owen et al., 1986). The effects produced thus range from useful anticonvulsant action (phenobarbital) to the deep narcosis produced by phenobarbital, thiopental, and other general anaesthetics. The fact that practically all general anaesthetics potentiate the GABA$_A$ inhibitory action (including gaseous agents such as halothane and isoflurane, and even steroid anaesthetics related to progesterone (Scholfield, 1980; Majewska et al., 1986; Cottrell et al., 1987) strongly suggests that enhanced GABAergic inhibition may be a significant component of the anaesthetic state. Of course, all these agents at anaesthetic doses have other actions—for example, they depress transmitter release and membrane Ca currents, and generally tend to

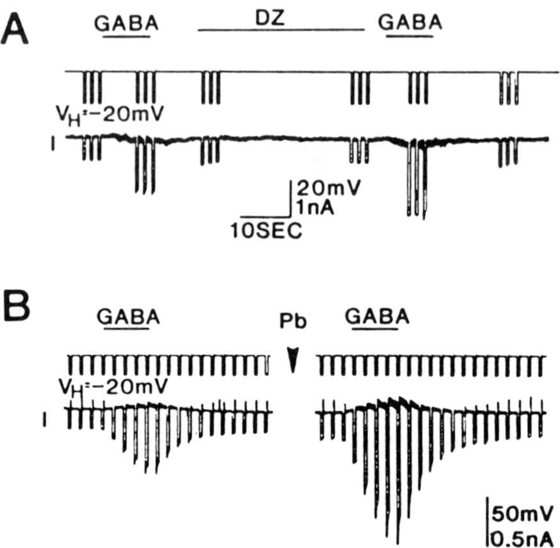

Fig. 7.   Diazepam and pentobarbital enhance GABA-evoked conductance of cultured rat hippocampal neurons. **A**: cell was voltage clamped at -20 mV, the reversal potential for the current response, and 20 mV hyperpolarizing commands were applied to estimate resting conductance. GABA was applied to the cell under control conditions and immediately after pressure application of 10 µM diazepam. Diazepam (DZ) enhances the GABA-activated conductance by about 50% while having little effect of its own on resting conductance. **B**: in another cell GABA-activated conductance (left) is markedly enhanced by 100 µM pentobarbital (Pb) (right). Each conductance response is associated with an increase in current variance and this is enhanced during the drug response. (Reproduced from Fig. 9 in Segal and Barker, 1984, with permission of the publisher.)

reduce electrical excitability (Nicoll and Madison, 1982) as well as excitatory transmitter actions (Macdonald et al., 1986). It is interesting to remember that several decades ago, Pavlov (1927) described the loss of awareness in sleep as due to a process of internal inhibition.

Modulators: Benzodiazepine site.   This site is perhaps of even greater potential value as a target for useful drug therapy, because the benzodiazepine receptor can bind a variety of agents capable of acting as selective agonists, antagonists as well as inverse agonists (Möhler and Okada, 1977; Squires et al., 1979; Olsen, 1982; Haefely et al., 1981; Polc et al., 1982; Macdonald et al., 1986; Owen et al., 1986; Martin, 1987). Agonists potentiate GABA (Figs. 7 and 8) and inverse agonists reduce its action; the situation is further complicated by the existence of drugs that act as *partial* agonists or *partial* inverse agonists (Polc, 1988; Haefely, 1989). The benzodiazapine receptor differs from the barbiturate-binding site not only in its pharmacological properties but also in that the potentiation of the GABA-evoked currents occurs in a different

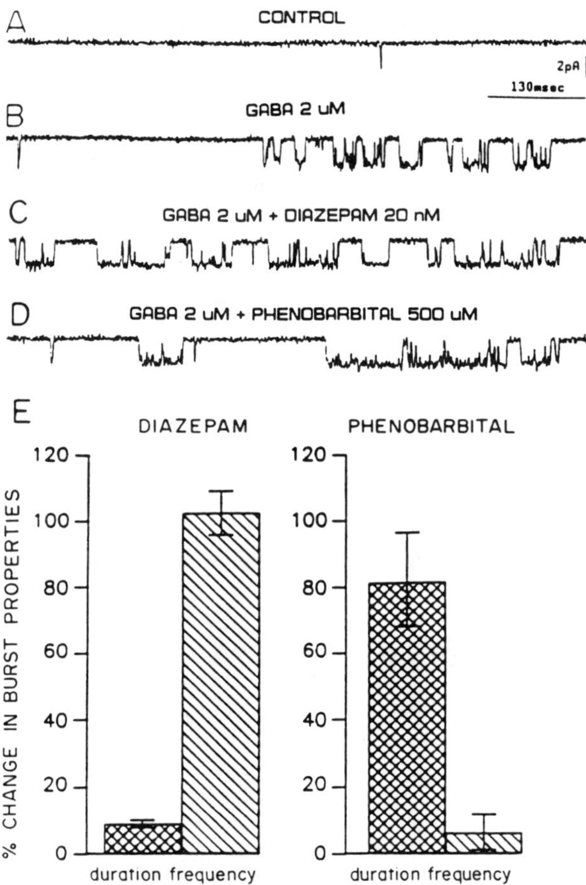

Fig. 8. Single channel chloride current evoked by GABA was enhanced by diazepam and phenobarbital. **A:** Before application of GABA, only rare brief spontaneous currents were present. **B:** Following application of GABA, channel openings were more frequent and occurred as individual openings or in groups of openings (bursts). **C:** In the presence of diazepam, GABA increased channel activity. **D:** In the presence of phenobarbital GABA again increased GABA receptor channel activity. Membrane patches were voltage clamped at the potassium equilibrium potential (-75 mV). Downward deflections in all tracings represent channel openings and inward chloride currents. **E:** histograms showing different effects of diazepam and phenobarbital on the properties of GABA evoked bursts. In the presence of diazepam, GABA-evoked bursts were more frequent but were not prolonged. In contrast, in the presence of phenobarbital, burst *frequency* was not changed, but burst *duration* was prolonged. (Reproduced from Figs. 2 and 6 in Twyman et al., 1989, with permission of the publisher.)

manner, by increasing the *frequency* of channel openings rather than their *duration* (Owen et al., 1986; Twyman et al., 1989) (Fig. 8).

An important reason for the great interest in benzodiazepines is the striking anxiolytic effect of agonist drugs; conversely, specific antagonists or inverse agonists have a powerful anxiety-promoting action. One therefore has to consider the possibility that anxiety levels are set by the relative amounts of endogenous agonists, antagonists or inverse agonists binding to this receptor; but in spite of an intensive search for such compounds, there is so far no really convincing evidence that they occur in the brain.

Not all effects of benzodiazepines should necessarily be ascribed to potentiation of GABA$_A$ actions (Polc, 1988); in particular, their marked depressant effect on adenosine uptake may well contribute to their tranquilizing properties (Phillis and Wu, 1980).

Thus, of the various modulators, only steroids, including progesterone and its derivatives (such as pregnanediol) are unquestionably endogenous agents,varying levels of which could significantly affect CNS function and behaviour (Selye, 1941; Majewska et al., 1986; Cottrell et al., 1987; Schwartz-Giblin et al., 1988).

GABA$_A$ receptor structure. Through the application of molecular techniques, studies of the structure and properties of the GABA$_A$ receptor have been advancing rapidly. For example, by injecting brain mRNA into oocytes, it became possible to have GABA$_A$ receptors expressed in a membrane that is normally not sensitive to GABA (Miledi et al., 1982; Sumikawa et al., 1984; Blair et al., 1988). The receptor itself has been purified from brain homogenates by benzodiazepine affinity chromotography (Sato and Neale, 1987; Schofield et al., 1987).

The isolated receptor was shown to consist of two polypeptide subunits (α and β, both of 50,000 Dalton), the first (α) having the benzodiazepine binding site, and the second (β) the GABA binding site (Fig. 9). Using appropriate peptide sequences, DNAs have been cloned which encoded both subunits, and could then be used to work out their amino acid sequences. To a high degree these proved to be homologous with each other; and more surprisingly with the nicotinic ACh receptor as well as the glycine receptor. This suggested that there is a general family ("superfamily") of directly ligand-gated channels, having a similar fundamental organization, which appears to differ markedly from that of the indirectly-gated channels that are G-protein dependent.

The extent to which one or the other subunit (and further α-subsubtypes) alone can reconstitute truly functional GABA$_A$ receptors, with all their characteristic pharmacological properties (including modulation by benzodiazepines and barbiturates) has been investigated systematically by cDNA or mRNA injections into Xenopus oocytes (Schofield et al., 1987; Blair et al., 1988) or by transfection into "naive" mammalian (kidney) cells (Pritchett et al., 1988). In view of the heterogeneity of the α-subtype especially (Schofield et al., 1987; Blair et al., 1988), selective probes are now being used to examine GABA$_A$ receptors in the brain by in situ hybridization, in order to ascertain whether different receptor subtypes are selectively expressed in various

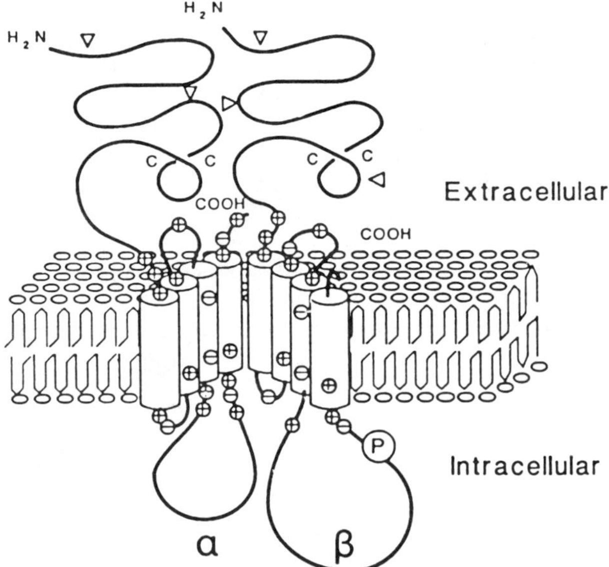

Fig. 9. A schematic model for the topology of the GABA$_A$ receptor in the cell membrane. Four membrane-spanning helices in each subunit are shown as cylinders. The structure in the extracellular domain is drawn in an arbitrary manner, but the presumed β-loop formed by the disulphide bond predicted at cysteines 139 and 153 is shown. Potential extracellular sites for N-glycosylation are indicated by triangles. Those charged residues which are located within or close to the ends of the membrane domains are indicated as small circles with charge marked. The site for cAMP-dependent serine phosphorylation, present only in the β-subunit, is marked by an encircled P. It is proposed that two copies of each of these subunit structures are complexed in the receptor molecule so as to align the membrane-spanning domains, only some of which will form the inner wall of a central ion channel. (Reproduced from Fig. 5 in Schofield et al., 1987, with permission of the publisher.)

regions of the brain. Another feature of interest is the presence of a serine site on the β subunit capable of being phosphorylated. This is especially relevant in the light of recent evidence that the GABA$_A$ receptor requires phosphorylation for its functional integrity (see page 68).

GABA$_B$ Receptors

That GABA might act on another kind of receptor should perhaps have been suggested by two features that have been known for a long time: one is the resistance of some GABA actions to bicuculline (Godfraind et al., 1970); and the second, the very different properties of the close GABA-derivative baclofen (β-p-chlorophenyl GABA) (Pierau and Zimmerman, 1973; Fox et al., 1978). But the first serious proposal that there exist two distinct populations of GABA receptors was made by Bowery et al. (1980; 1984). This idea is now widely accepted (e.g., Bormann, 1988; Bowery et al., 1989). Much evidence indicates two principal sites and mechanisms of GABA$_B$ action.

Presynaptic GABA$_B$ actions. A potent block of synaptic transmission in the spinal cord—affecting especially the monosynaptic reflex and entirely mediated at a presynaptic site—was strongly indicated by the experiments of Pierau and Zimmerman (1973) and Fox et al. (1978) (confirmed in recent studies, Peng and Frank, 1989; Edwards et al., 1989). This striking block is obtained with very low tissue concentrations of baclofen (0.1 µM), especially when they are compared with the 30–100 µM concentrations typically applied at other sites where baclofen suppresses the release of a variety of transmitters (Bowery et al., 1980). In low concentration, baclofen selectively inhibits the release of excitatory amino acids from cortical slices (Fig. 10) (Potashner, 1979; Potashner and Gerard, 1983).

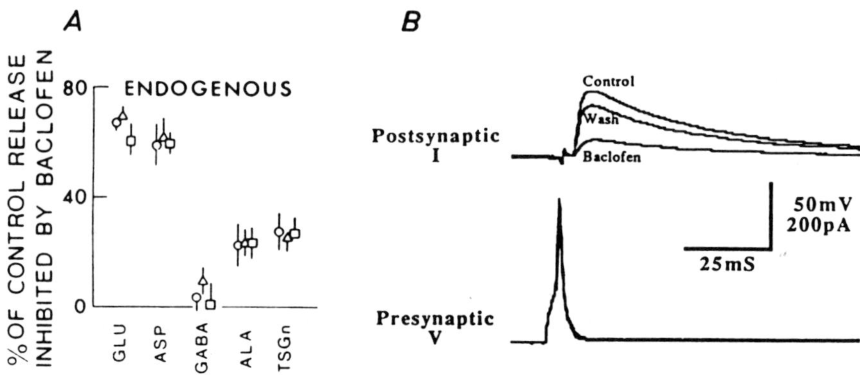

Fig. 10. A. Inhibitory effects of baclofen on electrically-stimulated release of amino-acids from cortical slices (**A**) and on IPSPs of hippocampal cell in culture (**B**). A: the symbols represent different estimates of the stimulated release. (TSG$_n$ = threonine, serine, glutamine). Inhibition was defined as the difference between the mean control stimulated release and the stimulated release in the present of baclofen, expressed as a percent of the former. The data are means ± SEM of the inhibitions observed in three experiments when 4 µM baclofen was present. Mean control stimulated releases were calculated from the results of 5–9 experiments in the absence of the drug (from Fig. 2 in Potashner, 1979). **B**: L(-)-Baclofen depresses monosynaptic inhibitory transmission. A single presynaptic action potential (below) evoked an outward-going monosynaptic IPSC (above). The postsynaptic cell was voltage-clamped at -40 mV. Each trace represents the average of 10–20 events. Application of 10 µM (-)-baclofen reversibly reduced IPSC amplitude. Note that the presynaptic action potential is unaltered; presynaptic input resistance was also unaffected (not shown). (Reproduced from Fig. 2 in Harrison et al., 1988, with permission of the publisher.)

This presynaptic block of transmission is not confined to the muscle afferent synapses on motoneurons: for example it is evident (at the same doses) at synapses made on motoneurons by fibres descending from the brainstem (Edwards et al., 1989) and at primary cutaneous afferent synapses in the cuneate nucleus (Fig. 11) (Fox et al., 1978). Both in the spinal cord (Davidoff and Sears, 1974) and in the cuneate (Fox et al., 1978) the presynaptic action of baclofen is associated with a reduction in terminal excitability (visible in Fig. 11B). In more recent experiments on dorsal root ganglion cells,

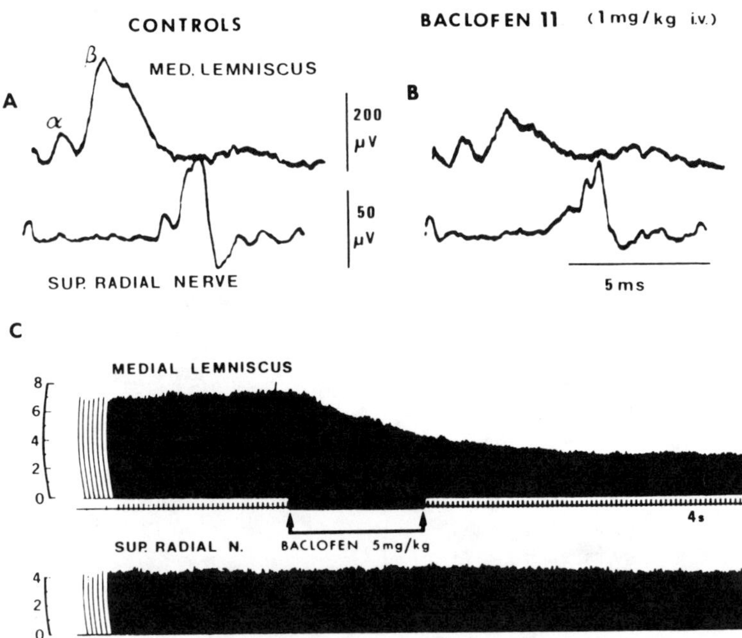

Fig. 11. Block of cuneate synaptic transmission (in cat) by intravenous baclofen. A, B: data showing that block is not accompanied by depression of postsynaptic electrical excitability. The stimulating microelectrode was inserted into the cuneate nucleus to a depth of 1.2 mm: hence the medial lemniscal response showed a very early α-wave due to direct stimulation of cuneothalamic neurones (upper traces). As usual, excitation of the afferent terminals evoked an antidromic volley in afferent fibres, recorded here in the superficial radial nerve (lower traces). A: controls; B: responses evoked by identical stimulation 11 min after injection of baclofen. C: in another experiment, time course of block of transmission through cuneate nucleus during and after intravenous injection of baclofen. Upper trace, medial lemniscal responses evoked by constant stimulation of afferent terminals in cuneate nucleus (at 1/s); middle trace: corresponding antidromic volley recorded in superficial radial nerve. Monophasic potentials were integrated and then recorded on the chart paper; scales are arbitrary. Arrows signal period of slow i.v. injection of baclofen. (Reproduced from Figs. 7 and 10 in Fox et al., 1978, with permission of the publisher.)

baclofen or GABA (acting via bicuculline-insensitive receptors, presumably of the $GABA_A$ type) were found to block Ca-currents (Dunlap and Fischbach, 1978; Bormann, 1988) by a G-protein mediated action (Dolphin and Scott, 1987). Whether this block of somatic Ca currents is relevant for the strong block of transmitter *release* produced by very low doses of baclofen in the spinal cord, the cuneate nucleus (Pierau and Zimmermann, 1973; Fox et al., 1978) or the hippocampus (Olpe et al., 1982; Inoue et al., 1985b) remains to be established. There is some evidence that spinal baclofen receptors may be distinct from GABA receptors (Sawynok, 1986). In contrast to the studies on amino acid release from cortical slices (Potashner, 1979), in which baclofen produced no change in GABA release, in electrophysiological experiments on hippocampal and cortical cells the release of GABA appears to be regulated by $GABA_B$ receptors (Misgeld et al., 1982; Harrison et al., 1988; Deisz and Prince, 1989; Thompson and Gähwiler 1989b; Davies et al., 1990). The reason for this discrepancy is not clear.

Post-synaptic $GABA_B$ actions. In addition to the longer-known pre-synaptic actions, baclofen and GABA (acting via bicuculline-resistant receptors) have a significant post-synaptic action, mediated by G_K (not G_{cl}) increase (Gähwiler and Brown, 1985; Inoue et al., 1985a; Newberry and Nicoll 1985; Peet and McLennan 1986; Osmanovíc and Shefner, 1988). The G_K-activation is G-protein dependent (Asano et al., 1985; Andrade et al., 1988; Dutar and Nicoll, 1988b), possibly sharing the same G_K channels with other G-protein dependent transmitter actions (such as that of 5-HT). It probably accounts for the late component of GABAergic IPSPs recorded in the hippocampus (Newberry and Nicoll, 1985; Dutar and Nicoll, 1988a; Thalmann, 1988; Lambert et al., 1989) and neocortex (Connors et al., 1988; McCormick 1989) (Fig. 12).

The presynaptic block of EPSPs by baclofen in the hippocampus differs from the post-synaptic activation of G_K in being *insensitive* to either the $GABA_B$ receptor antagonist phaclofen or to the G-protein inhibitor pertussis toxin (Dutar and Nicoll, 1988a)—this suggests possibly different receptors and/or coupling mechanisms. A further complication is that $GABA_B$ receptor agonists enhance cyclic AMP formation induced in the forebrain by catecholamines and some other agents, although not in other regions of the CNS (Karbon and Enna, 1985).

The lack of powerful and specific, slowly-reversible ligands has so far precluded any successful attempts at *isolating* the $GABA_B$ receptors. No modulation (potentiation) of $GABA_B$ receptors has yet been reported.

GABA Uptake

GABA is very effectively removed from the extracellular space by uptake into brain cells, as originally found by Elliot and van Gelder (1958) and repeatedly since (Henn and Hamburger, 1971; Krogsgaard-Larsen, 1980; Dingledine et al., 1988). As might be expected, the uptake mechanism does not differ from other amino acid (and sugar) transport in being linked to Na^+ movements along the Na^+ electrochemical gradient, and therefore

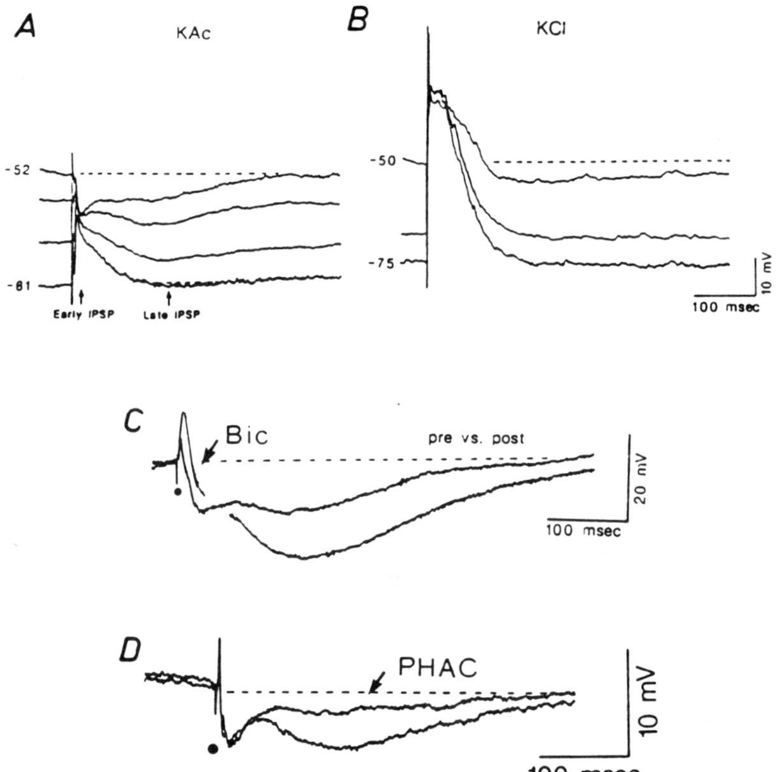

Fig. 12. GABA$_A$ and GABA$_B$ receptor-mediated components of IPSPs in slices of human cerebral cortex. Effect of intracellular injection of Cl$^-$ on evoked IPSPs. **A:** Normal PSP sequence evoked by electrical stimulation of afferents in a neuron recorded with a potassium acetate (KAc)-filled microelectrode. An initial EPSP is followed by both an early IPSP and a later IPSP. Early IPSP reverses at about -75 mV. **B:** Afferent stimulation in a neuron recorded from the same region as the cell in A, with a microelectrode filled with potassium chloride (KCl) results in a large and prolonged depolarization followed by a later hyperpolarization. **C:** Superimposed traces show that GABA$_A$ antagonist bicuculline (Bic) selectively reduces the early IPSP. Single shock stimuli (.) give rise to an EPSP followed by 2 phases of IPSP; local application of bicuculline (25 μM in micropipette) quickly (1 min) results in a potentiation of the EPSP and late IPSP and a diminution of the *early* IPSP. Membrane potential held at -63 mV; resting membrane potential was -79 mV. **D:** GABA$_B$ antagonist phaclofen (PHAC) specifically reduces the later IPSP. Local application of phaclofen (5 mM in micropipette) selectively and reversibly reduces the later IPSP. Membrane potential held at -58 mV. Resting membrane potential was -72 mV. All data obtained from layer II-III cells. (Reproduced from Figs. 2, 3, and 5 in McCormick, 1989, with permission of the publisher.)

ultimately depends on the activity of the Na/K pump (Schultz and Curran, 1970; Erecinska, 1987). The efficacy of GABA uptake in keeping a low extracellular GABA concentration, and its crucial dependence on the Na gradient is well illustrated by Akaike, Maruyama et al.'s (1987) ingenious experiment (Fig. 13). The potency of GABA was strongly enhanced by

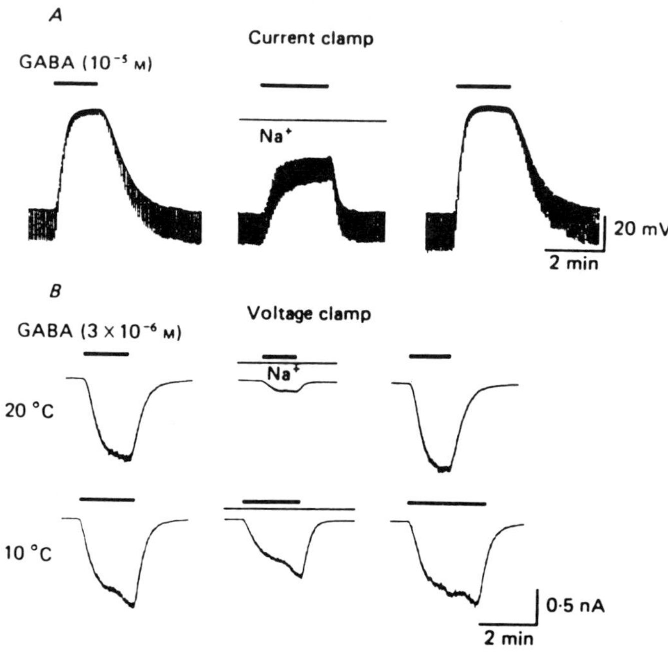

Fig. 13. In frog sensory neurons, Na^+-driven, temperature-sensitive GABA transport greatly attenuates GABA's action. **A:** decrease in the GABA responses with $[Na^+]_0 = 90$ mM. During control runs, external Na^+ was substituted by $Tris^+$. Both Na^+ and $Tris^+$ external solutions contained 120 mM Cl^-. The cell was perfused with K^+- and Ca^{2+}-free internal solution containing 9 mM-Na^+ and 120 mM-Cl^-, under conditions of current clamp. The thick horizontal bar above each response indicates the period of bath application of 10^{-5} M-GABA; while the thin bar indicates the interval during which the Na^+-containing external solution was applied. The membrane potential was held at -60 mV by d.c. current injection, before initiation of GABA application. To measure the input resistance, hyperpolarizing current pulses (0.5 nA, 400 ms) were applied through a suction pipette every 4 s. The brief voltage shift corresponds to each pulse. The depolarizing GABA response decreased with $[Na^+]_0 = 90$ mM. **B:** effects of cooling on 3×10^{-6} M-GABA-induced inward I_{cl} under voltage clamp with a holding potential (V_H) of -60 mV; note much smaller current is induced by GABA with $[Na^+]_0 = 90$ mM, compared with the controls ($Tris^+$) in the same neurone. Cooling the ganglion reduced the difference between the test and control results. (Reproduced from Fig. 1 in Akaike, Maruyama et al., 1987, with permission of the publisher.)

removal of Na_0 (or increase in Na_i); by reducing the Na^+ gradient, the rapid uptake of applied GABA was prevented.

Although GABA uptake may play only a minor role in limiting the duration of IPSPs (only the later part of which seems to be affected by blockers of uptake (Dalkara, 1986; Korn and Dingeldine, 1986; though cf. Rekling et al., 1990), its functional importance has probably been underrated. The following points need to be kept in mind:

1. Because GABA transport is acutely dependent on the transmembrane Na electrochemical gradient, its efficiency is accordingly much diminished when cells are depolarized.

2. Like all such transport mechanisms, GABA uptake is reversible. Depolarization (e.g., by high $[K]_0$) should cause GABA efflux. Thus when high K is applied (a standard procedure in studies of GABA release) much of the release seen must take place by reversed GABA transport, rather than via the Ca-dependent (presumably vesicular) mechanism (Erecińska, 1987; Nicholls, 1989).

3. Increases in $[K]_0$ can of course be very substantial, to as much as 10–12 mM in cortex and hippocampus during seizure activity (Somjen, 1979). But in the hippocampus *in situ*, they are surprisingly large even during very low frequency, non-epileptiform activities (Krnjević et al., 1980) and are associated with correspondingly large glial depolarizations (Casullo and Krnjević, 1987). The resulting GABA release could be a significant component of a local homeostatic mechanism, tending to limit over-depolarization by a non-synaptically mediated inhibition. Indeed, GABA may be released in this way from horizontal cells in the retina (Schwartz, 1987).

4. During hypoxia progressive terminal depolarization initiates transmitter release, mainly by this Ca-independent mechanism (Nicholls, 1989). Of course, other transmitters are released as well as GABA, notably glutamate (Benveniste et al., 1984); but to the extent that GABA is released, it would have a protective action by limiting the rate of depolarization. However, if depolarization is excessive and spreading depression occurs, one can expect a massive release of all the stored transmitters, including GABA. This presumably contributes to the short-circuiting of membrane resistance observed during spreading depression (Snow et al., 1983; Leblond and Krnjević, 1989).

5. GABA itself has no net charge near pH 7. Its transport is Na^+ linked (Erecińska, 1987) and therefore electrogenic. Associated movements of Cl^- have also been described (Kanner and Radian, 1985); but these are more likely to be secondary to the Na^+ movement (Erecińska, 1987). Thus, GABA uptake can be expected to have a depolarizing influence. Whether this would be significant depends on the rate of transport and on the input resistance. The depolarizing effect of GABA on glia in situ was tentatively ascribed to such a mechanism (Krnjević and Schwartz, 1967); but according to much more recent studies (Kettenmann et al., 1987), the depolarization of cultured glial cells appears to be mediated by activation of $GABA_A$ receptors and Cl-channels. Evidence that $GABA_A$ receptors are involved in the depolarizing effect of GABA on glia in hippocampal slices (MacVicar et al., 1989) is less convincing, because extensive electrical coupling precludes reliable

measurements of conductance changes or reversal potential. If indeed GABA activates glial $GABA_A$ receptors, Cl^- release from the relatively Cl-rich glia could help to restore $[Cl]_0$ depleted by Cl^- influx into nerve cells during GABAergic IPSPs (as suggested by MacVicar et al., 1989).

6. A final point to bear in mind is that because GABA is involved in so many inhibitory mechanisms, even a minor depression of GABA uptake may have major functional consequences. Thus systemic administration of a blood-brain barrier permeable GABA-uptake inhibitor (Ebert and Krnjević, 1990) recently proved to have quite dramatic and long-lasting behavioural and psychogenic effects in human subjects (Taylor and Vartanian, 1989).

LABILITY OF GABAERGIC INHIBITION

Here we consider a curious paradox. By all account, GABA causes a powerful inhibition, especially through the $GABA_A$ and to a lesser extent $GABA_B$ receptors (e.g., Connors et al., 1988; McCormick, 1989). GABAergic cells and synapses are found throughout the brain. Any kind of discrete organized function would probably be impossible in the absence of GABA (see page 77). And yet, GABAergic inhibition is surprisingly labile.

In view of their vital importance, IPSPs might be expected to show high resistance to adverse conditions. In fact, the opposite seems to be the case: repetitive stimulation in the hippocampus soon leads to a sharp diminution of IPSPs (Fig. 14) (Ben-Ari et al., 1979), and, in the absence of strong inhibition, a rapid build-up of excitation resulting in "burst discharges". What can be the explanation for this unexpected lability of IPSPs?

Mechanisms of IPSP Failure During Repetitive Stimulation

This phenomenon has been extensively studied in the hippocampus and neocortex, by several different groups (Wong and Watkins, 1982; Numann and Wong, 1984; McCarren and Alger, 1985; Stelzer et al., 1987; Deisz and Prince, 1989; Thompson and Gähwiler 1989a,b). So there is now much relevant information; however, all of it has been obtained from isolated preparations in vitro, and it is not certain whether it is fully applicable to the original observations (Ben-Ari et al., 1979), made in the hippocampus in the brain in situ. (Though extremely convenient, slices differ significantly from brain tissue in situ, especially with regard to their metabolic status (Lipton and Whittingham, 1984)). At any rate, a certain consensus seems to be emerging from the above studies in favour of a presynaptic mechanism—a depression of transmitter (GABA) release — rather than a postsynaptic one, such as desensitization.

Loss of GABA sensitivity. A "fading" of the conductance increase caused by maintained applications of GABA is a well-known characteristic feature (Dreifuss et al., 1969; Krnjević, 1984). Unlike the reduction in the *amplitude* of IPSPs, this could not be explained simply by a change in Cl^- gradient resulting from the GABA-induced influx (a positive shift in V_{cl} if anything increases G_{IPSP} (cf. Thompson and Gähwiler, 1989a) and depolar-

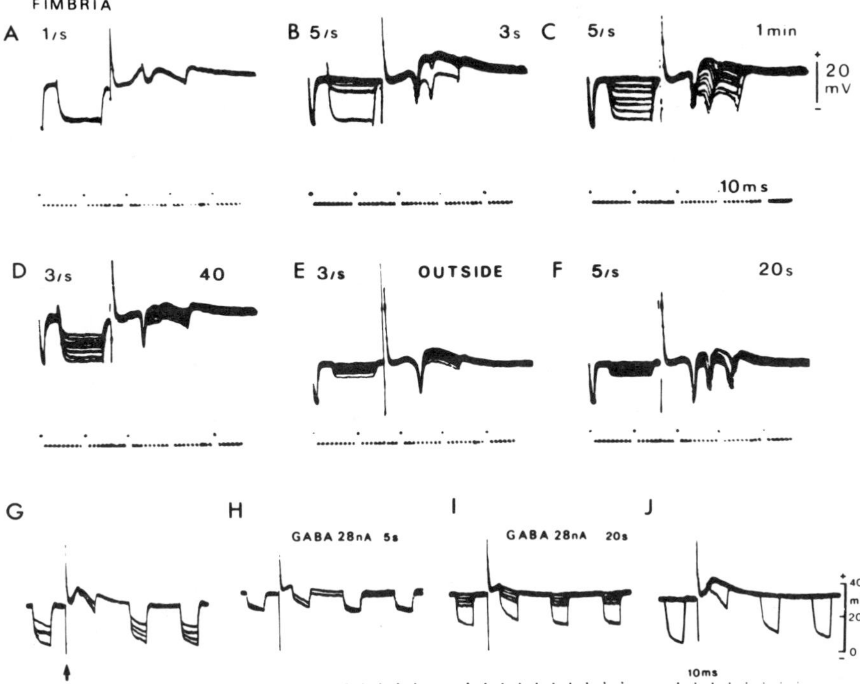

Fig. 14. Lability of inhibition and GABA action in rat hippocampus (in situ). **A:** Large IPSP conductance increase was evoked by constant fimbrial stimulation. IPSP conductance change started to diminish within 3 sec of raising frequency of stimulation from 1 to 5 Hz (**B**) and it practically disappeared within 1 min (**C**). Recovery was almost complete 40 sec after reducing stimulation frequency to 3 Hz (**D**). Traces **E** and **F** are controls recorded outside the cell during 3- and 5-Hz stimulation. Traces **G-J** illustrate GABA-evoked conductance increase and its rapid "fading". (**G**), Control record with high conductance at peak of IPSP; in **H**, 5 sec after start of GABA iontophoresis, there is a large fall in resting resistance (cf. first pulse at left), but 15 sec later, GABA effect is greatly diminished, even though iontophoretic release was fully maintained: **J** shows recovery of both resting resistance and IPSP conductance change 1 min after end of GABA application. (Reproduced from Fig. 9 in Krnjević, 1983.)

ization also enhances both G_{IPSP} and G_{GABA} (Gray and Johnston, 1985; Weiss et al., 1988; McCormick, 1989)).

One possible mechanism of fading would be GABA uptake, if this can significantly lower the concentration of GABA at the receptors (Krnjević, 1984). This is not a purely speculative notion, because fading can be strikingly reduced or even abolished by an inhibitor of GABA uptake, such as nipecotic acid (Fig. 15) (Rovira et al., 1984; Dalkara, 1986). However,

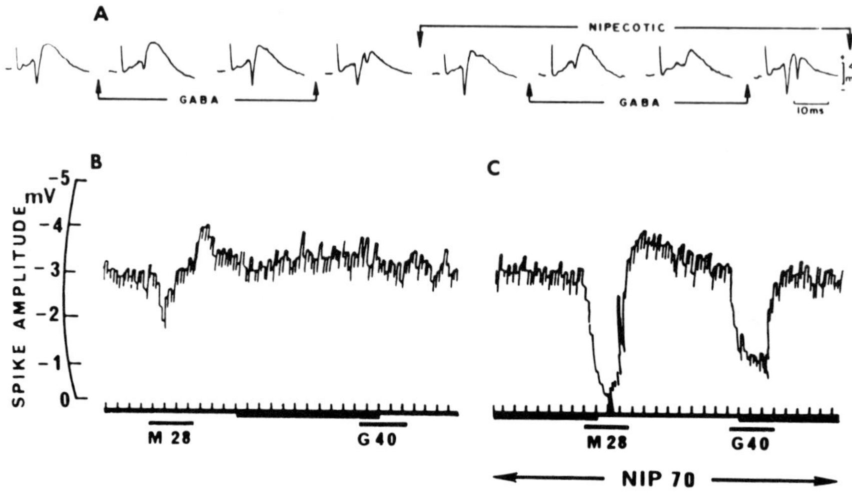

Fig. 15. **A:** GABA-uptake markedly attenuates inhibitory effect of GABA on CA1 stratum pyramidale population spikes, in rat hippocampus in situ. This is indicated by strong potentiation of GABA action (and disappearance of fading) during concurrent-iontophoretic release of the uptake inhibitor, nipecotic acid. Oscilloscope traces of the population spike evoked by 6 V fimbrial stimulation. **B, C:** Slow chart-paper records of the peak amplitude of population spike evoked by constant fimbrial stimulation; time marks indicate 5 s intervals. In **B**, 28 nA of muscimol depresses the spike; however, that effect fades quickly and is followed by a rebound. GABA (40 nA) has a hardly detectable inhibitory action. **C:** With concomitant release of nipecotic acid (70 nA), muscimol (M) and GABA inhibit the spike strongly and there is little or no fading (Fig. 2 in Dalkara 1986).

when applied during repetitive stimulation nipecotic acid, not only did not prevent the failure of IPSPs but indeed accelerated the phenomenon (Deisz and Prince, 1989). This was attributed to the increased extracellular concentration of GABA. But of course, since nipecotic acid itself is preferentially taken up instead of GABA (Johnston et al., 1976; Ryan and Schwartz, 1986), it may tend to displace GABA from the release sites (and indeed may be released itself as a false transmitter) thus greatly diminishing GABA release. So this is a not very useful test.

A more conventional mechanism of desensitization would be through some secondary conformational change, of the kind originally postulated by Katz and Thesleff (1957) and likely to be responsible for the rapid desensitization seen in patch clamp studies of GABA (Hamill et al., 1983; Akaike, Inoue et al., 1987; Bormann et al., 1987). As already mentioned, if GABA uptake is prevented, fading may vanish altogether: under these conditions (in situ), desensitization is not conspicuous, perhaps because it depends critically on the GABA concentration; or because some factor is present in situ that opposes desensitization; alternatively, it may happen too quickly to be readily noticeable.

An important cause of GABA desensitization is probably a rise in intracellular Ca^{2+}. This was first demonstrated by Inoue et al. (1986) when they found that dorsal root ganglion cells' responses to GABA were markedly suppressed by Ca^{2+} influx (Fig. 16). This mechanism could explain a progressive reduction of IPSPs (and GABA responses) by repetitive stimulation that activates Ca currents (Krnjević et al., 1986; Connor et al., 1988). Indeed Stelzer et al. (1987) observed a loss of IPSPs and GABA responses following repeated activation of N-methyl-D-aspartate (NMDA) receptors during kindling (Fig. 17). In later experiments, Stelzer et al. (1988) found that GABA$_A$ receptors are also dependent on continuing phosphorylation (possibly at the β-subunit site identified by Schofield et al., 1987 cf. Fig. 9). A diminished sensitivity to GABA during and after tetanic stimulation has often been observed (Ben-Ari et al., 1979; Wong and Watkins 1982; Stelzer et al., 1987; Kapur et al., 1989; Thompson and Gähwiler, 1989b), but it probably accounts for only part of the depression of IPSPs. A reduction of

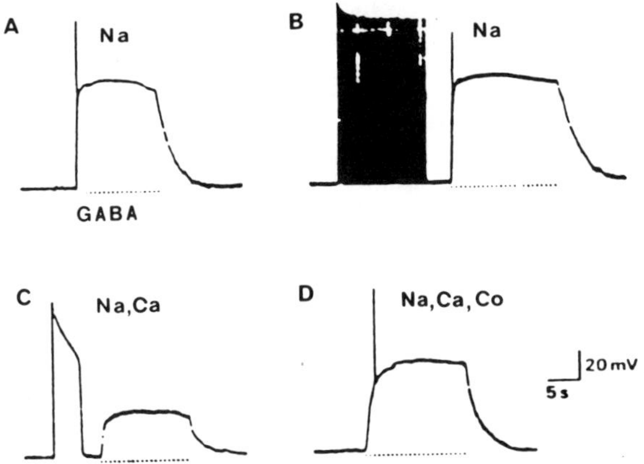

Fig. 16. Ca-influx into frog dorsal root ganglion cell greatly depresses responses to GABA. Recording from freshly dissociated cell was with suction pipette, in solution containing only Cl salts of Na, Mg, TEA and Cs (**A,B**), or with Ca instead of Mg (**C,D**). Large depolarization evoked by 10 µM GABA (**A**) was not significantly changed by preceding firing (50 depolarizing pulses at 3.3 Hz) that generated only Na$^+$ spikes (**B**). But a preceding long Ca^{2+} spike sharply reduced the effect of GABA (**C**). When the Ca^{2+} spike was eliminated by 4 mM Co^{2+}, GABA again caused a marked depolarization (**D**); this showed that *external* Ca^{2+} does not interfere with the action of GABA. (Reproduced from Fig. 1 in Inoue et al., 1986, with permission of the publisher.)

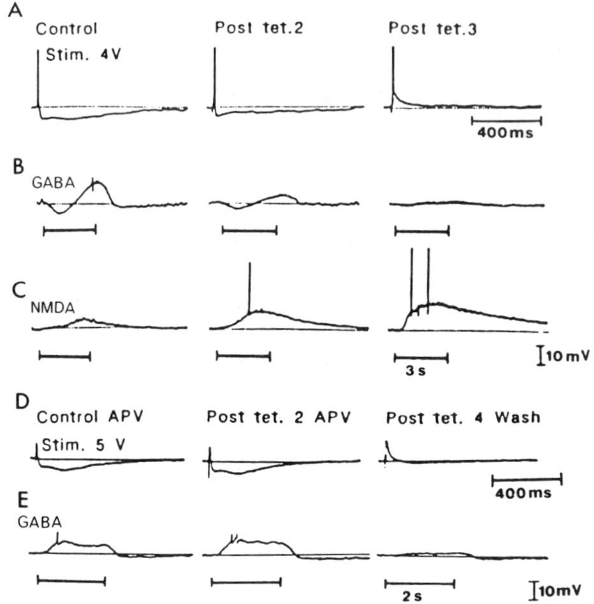

Fig. 17. Post-tetanic depression of IPSPs is accompanied by loss of responses to GABA and increased responses to NMDA: these effects of repeated tetanic stimulations can be prevented by D-aminophosphonovalerate (APV), an NMDA antagonist. In hippocampal slice GABA and NMDA were applied to CA1 neuron iontophoretically (by 11 and 22 nA respectively) and traces were obtained at constant membrane potential. Each tetanic stimulation consisted of 5 trains (50 Hz for 2s) repeated at 10 s intervals, and was applied regularly every 30 min. **A-C:** near-simultaneous recordings show progressive and parallel post-tetanic depression of IPSPs and GABA effects, and increase in NMDA effects. **D, E:** in presence of 5–10 µM APV, the same tetanic stimulation failed to depress either IPSP or the effect of GABA (52 nA), but the usual post-tetanic depressions were seen after washing out APV. (Reproduced from Fig. 4 in Stelzer et al., 1987, with permission of the publisher.)

GABA release appears to play a (perhaps *the*) major role in the tetanic suppression of IPSPs.

Failure of GABA release from GABAergic terminals. It is not likely that GABA would be seriously depleted at the relatively low frequency of stimulation used in many experiments (< 10 Hz). More significant could be local ionic changes (Krnjević et al., 1980; Marciani et al., 1982). $[K]_0$ can easily reach a level of > 10 mM, at which IPSPs vanish (cf. Thompson and Gähwiler, 1989a), most likely owing to a failure of terminal invasion (Krnjević and Miledi, 1959; Swadlow et al., 1980). $[Ca]_0$ on the other hand drops sharply, especially in the pyramidal cell layer (Krnjević et al., 1980; Marciani et al., 1982) where inhibitory terminals are especially concentrated. This could cause a selective disinhibition. Tetanic stimulation probably also releases agents that are known to depress transmitter release: for example, ACh which has a marked disinhibitory action in the hippocampus (Ben-Ari et al., 1981b; Haas, 1982; Segal, 1983).

The importance of GABA release has been emphasized by several groups of investigators. Though long suspected (by analogy with spinal presynaptic inhibition, where a picrotoxin-sensitive GABAergic mechanism was recognized very early (Eccles, 1964)), a negative feedback action of released GABA has been convincingly demonstrated only quite recently. Both in neocortex (Deisz and Prince, 1989) and hippocampal cells in monolayers (Thompson and Gähwiler, 1989b), there is a powerful depression of IPSPs by baclofen or exogenous GABA, presumably acting through presynaptic GABA_B receptors. The clinching evidence has come from very recent experiments (Davies et al., 1990) on IPSPs evoked by direct stimulation of inhibitory neurons—that is in the absence of excitatory synaptic transmission, blocked by appropriate glutamate antagonists—which showed not only that such IPSPs could be suppressed by GABA and baclofen, but in addition that the depression of IPSPs seen with repeated stimulation could be reduced or abolished by the selective GABA_B antagonists phaclofen and 2-hydroxy-saclofen (Kerr et al., 1988). Although these experiments could not identify the precise locus of GABA autogenic inhibition, previous evidence (illustrated in Fig. 11B) indicated that baclofen acts only on the terminals of cultured inhibitory cells (and *not* their cell bodies) (Harrison et al., 1988). So it is most likely that the block of GABA release in hippocampal slices is mediated by GABA_B receptors situated on the GABAergic terminals in hippocampal slices. A significant feature of the cultured cells is the absence of any somatic GABA_B receptors (Fig. 11B). This provides further support for the idea that the presynaptic GABA_B receptors differ in some important way from the post-synaptic ones, notably in their non-dependence on a pertussis toxin sensitive G-protein (Dutar and Nicoll, 1988a). One can conclude at this point that, though a number of factors contribute to the rapid decay of hippocampal cortical IPSPs during repeated stimulation (including GABA "fading", Cl⁻ equilibration, and other ionic changes), the most important single factor may be autogenic inhibition of GABA release, mediated by GABA_B receptors situated on the GABAergic nerve terminals.

Significance of frequency-dependent failure of GABAergic IPSPs. Although we now have a better idea of how IPSPs fail during repetitive activation, the paradox remains; why should inhibition be so labile? The clue to the puzzle may lie in the extraordinary ability of the hippocampus to switch from a state characterized by very moderate activity to one in which powerful synchronized bursts predominate (e.g., Fig. 14E, F). A number of factors undoubtedly contribute to the large synchronized depolarizations and firing—such as activation of NMDA receptors, K⁺ accumulation and ephaptic interactions—but it may well be that the critical amount of depolarization needed to turn on NMDA receptors can be achieved only when both feedback and feedforward inhibitory synapses are significantly weakened (Dingledine et al., 1986; Collingridge et al., 1988). It is not clear in what way synchronized activity in bursts would be useful to the hippocampus (or the animal), especially as it may readily degenerate into full-blown seizures and/or spreading depression. Perhaps this is the price that has to be paid in order that hippocampal cells and circuits be able to respond to significant inputs with brief excitations of sufficient intensity to activate long-term storage mechanisms. A frequency-dependent failure of IPSPs may thus be important for the initiation of tetanically-induced long-term potentiation (LTP) (Davies et al., 1991). Whether a persistent depression of inhibitory synapses is also involved in the sustained facilitation (LTP) of population spikes, as opposed to EPSPs, is still not settled (Fig. 18) (cf. Haas and Rose, 1984; Griffith et al., 1986; Stelzer et al., 1987; Chavez-Noriega et al., 1989).

Failure of IPSPs during anoxia. Fujiwara et al. (1987) first pointed out that IPSPs in hippocampal slices are selectively abolished at an early stage of anoxia (see also Leblond and Krnjević, 1989). The mechanism of anoxic disinhibition has not been analyzed in detail. It is probably not due to a loss of GABA sensitivity—which might follow either a rise in $[Ca]_i$ or a

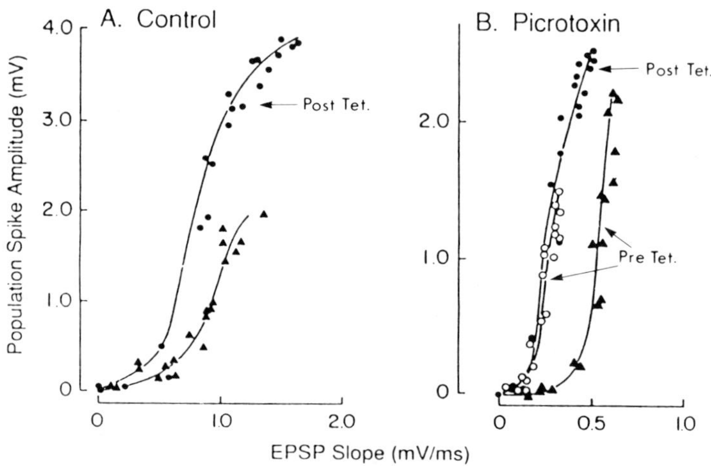

Fig. 18. Long-term potentiation (LTP) of EPSP-population spike (E-S) coupling disappears in the presence of picrotoxin: this suggests that LTP of E-S coupling is caused by sustained depression of GABAergic inhibition. In slices of rat hippocampus, extracellular recordings were made of CA1 field EPSPs and corresponding population spikes evoked by different intensities of stimulation. **A:** E-S curves from control slice show usual marked post-tetanic enhancement (triangles, initial data; circles, 30 min after high-frequency tetanus). **B:** in a different slice, comparable E-S curves were obtained first before (triangles) and then after (open circles) the introduction of picrotoxin (10 μM): note marked potentiation, as well as no further increase in slope after tetanic stimulation (closed circles). (Reproduced from Fig. 2 in Chavez-Noriega et al., 1989, with permission of the publisher.)

drop in ATP supply (cf. Stelzer et al., 1988), but rather a failure of GABA release, caused by selective inactivation of inhibitory cells or terminals (Krnjević et al., 1991). Whatever the mechanism, it appears to be relatively slowly reversible. A cumulative disinhibition may explain the progressive increase in hyperactivity seen in slices after repeated anoxic episodes (Ben-Ari and Cherubini, 1988).

The most conspicuous early effects of anoxia on nerve cells (such as hyperpolarization) tend to reduce firing and thus can be seen as protective responses. The block of IPSPs is clearly *not* of that kind. The greater susceptibility of inhibitory actions to brief anoxia is all the more surprising in view of the striking resistance of hippocampal GABAergic cells to irreversible ischemic damage (Pulsinelli, 1985).

CONCLUSIONS
Distribution of GABA$_A$ Receptors

In at least one respect, GABA can justly be considered a unique transmitter. Huge numbers of GABAergic cells are present in all regions of the CNS, and their terminals exert powerful effects on most cells. But the same can be said of glutamatergic cells. What is unique is the extraordinary variety of nerve cells that are inhibited by GABA.

Practically *all* central nerve cells (and perhaps glia) are sensitive to GABA: more surprising is the fact that many peripheral neurons, which are neither GABA-containing nor innervated by GABAergic terminals (such as cells in dorsal root and sympathetic ganglia) and even peripheral nerve fibres (with the possible exception of motor fibres, Morris et al., 1983) have functional GABA$_A$ receptors, for which there is no obvious function. Why are such cells provided with receptors for GABA (but *not* for glutamate or aspartate)? Is it because GABA$_A$ receptors are members of a ubiquitous family of receptors (Schofield, et al. 1987), whose basic elements are manufactured by all nerve cells, for use perhaps during development, or regeneration after injury, serving a GABA-mediated trophic action; or which are sufficiently interchangeable to act as receptors for ACh, GABA, glycine (or some other transmitter) as the need arises?

GABA$_A$ Receptor-Mediated Inhibition

So far, no consistent excitatory transmitter function has been ascribed to GABA. But its action is by no means universally hyperpolarizing. The great variability of $[Cl]_i$ (even in the same cell) and the significant HCO_3^- permeability of GABA$_A$-activated channels allow for much variation of V_{IPSP}. In spite of that, GABAergic action is overwhelmingly inhibitory, even when associated with substantial postsynaptic depolarization (as in spinal presynaptic inhibition). This can be explained by the very large increases in input conductance produced by GABAergic inhibition: by short-circuiting the membrane, GABA effectively uncouples the region of the cell where spikes are initiated from the dendritic sites where most EPSPs are generated. Of course much more discriminating local inhibitions can be exerted at discrete points in the dendritic tree.

Almost every imaginable kind of synaptic or even non-synaptic mechanism or arrangement seems to be present in the brain. GABA release can be Ca-dependent (presumably vesicular) and non-Ca-dependent (by reversed GABA uptake). The inhibition can operate in a feedback or feed-forward manner. A marvellously flexible inhibitory system thus provides for all possible kinds of control: from general reductions of excitability to inhibitions that are highly selective with regards to time and target. As if this were not enough, the whole system is subject to modulation in various ways: through local pH (and corresponding changes in HCO_3-); changes in $[Ca]_i$ or ATP levels; GABA uptake; countless drugs (and perhaps endogenous ligands) that bind at the benzodiazepine and barbiturate sites; and even its own rate of activation ("frequency-dependence") mediated by autogenic inhibition of GABA release.

Functional Organization of GABAergic Inhibition

Nearly all the evidence available with regard to the physiological role of GABAergic inhibition relates to $GABA_A$ receptor-mediated actions. The widespread distribution of baclofen (presumed $GABA_B$) receptors is attested by the marked pharmacological (and therapeutic) actions of baclofen. There is now good evidence that postsynaptic baclofen receptors are responsible for the later component of hippocampal and neocortical IPSPs (Connors et al., 1988; Dutar and Nicoll, 1988b; McCormick, 1989) and presynaptic ones for the frequency-dependent "fatigue" of IPSPs (Davies et al., 1990). (The general presumption that GABA is the transmitter actually involved in these actions is reasonable but not yet based on incontrovertible facts). To what extent normal function depends on baclofen ($GABA_B$)-receptor activation is still largely a matter of conjecture (cf. Connors et al., 1988) that should soon be resolved now that specific and increasingly powerful baclofen antagonists are becoming available (Kerr et al., 1987; 1988).

Disinhibition as a Principle of Functional Organization

In view of the huge number of inhibitory synapses and their great potency in regulating cell firing, the idea naturally came about that *action* in the brain was essentially generated by *disinhibition*. In other words, if most neurons are tonically inhibited—any tendency to firing in response to a given input being moreover automatically self-limited by widespread recurrent inhibition—then the brain forms a very stable system, unlikely to go spontaneously (or be triggered easily) into uncontrolled activity. On the other hand, purposeful discharges can be readily initiated by the selective and graded removal of local inhibitory action. This model has much in its favour, as discussed and analyzed in very plausible (if speculative) detail by Roberts (1976). It has not been superseded by the currently more fashionable speculations concerning the role of excitatory amino acids in "plasticity." As already discussed above, *disinhibition* may well be an essential permissive step or trigger for the enhanced synaptic excitatory action of glutamate that activates NMDA receptors (Muller et al., 1988; Davies et al., 1991).

Loss of Inhibition as a Principle of Functional Disorganization

There is now overwhelming evidence that a major loss of GABAergic inhibitory action in the brain leads to uncontrolled cellular discharges, resulting in electrographic seizures, often accompanied by motor seizures. As discussed in detail in previous reviews (e.g., Krnjević, 1983; Ribak, 1987), seizure activity is either initiated or facilitated by GABA antagonists (such as picrotoxin and penicillin), by drugs that block GABA synthesis (like semithiocarbazide), and by procedures or agents that selectively damage GABAergic neurons (in various epileptic models produced by focal applications of Co^{2+}, alumina cream, freezing, for example). Moreover, some of the most effective antiepileptic agents are drugs that reinforce GABAergic inhibition, like phenobarbital and diazepam (all these drugs are $GABA_A$ receptor ligands: so far there has been no evidence that baclofen antagonists have major convulsant actions and baclofen itself is not a useful anticonvulsant drug; it is therefore unlikely that loss of $GABA_B$ receptor action plays a crucial role in epileptic disorders). As already mentioned, GABAergic pathways through the substantia nigra seem to be involved in controlling the susceptibility to generalized seizures (McNamara et al., 1984; Gale, 1985).

It is only fair to point out, however, that such a widely used anti-epileptic drug as phenytoin is not known to have any relation to GABAergic mechanisms (Macdonald et al., 1985); and that serious doubts have been raised concerning a significant depression of inhibitory mechanisms in the penicillin model of generalized epilepsy (Giaretta et al., 1987). Clearly one cannot absolutely exclude other possible mechanisms of epileptogenesis—such as potentiation of excitatory synaptic or membrane mechanisms. Nevertheless, it is equally clear that a significant loss of GABAergic inhibition, whatever its mechanism, must tend to facilitate excessive discharges and therefore promote acute or chronic epileptiform activity.

Possible Long-Term Effects of GABA

In the present article—as in most discussions of GABAergic function—the main emphasis has been placed on GABA's uniquely widespread role as inhibitory transmitter. The notion that fast-acting transmitters are exclusively involved in rapid signalling is not consistent with the fact that even glutamate activates receptors that are not linked to ion channels (Sugiyama et al., 1987). Similarly, the modulation of cyclic AMP formation by $GABA_B$ receptors (Karbon and Enna, 1985) provides one mechanism (and there may be others) through which GABA could well play a significant role in longer term metabolic or trophic changes.

ACKNOWLEDGMENTS

The author's research is financially supported by the Medical Research Council of Canada.

REFERENCES

Akaike N, Maruyama T, Sikdar SK, Yasui S: Sodium-dependent suppression of γ-aminobutyric-acid-gated chloride currents in internally perfused frog sensory neurones. J Physiol (Lond) 392: 543–562, 1987.

Akaike N, Inoue, M, Tokutomi N: Contribution of chloride shifts to the fade of γ-aminobutyric acid-gated currents in frog dorsal root cells. J Physiol (Lond) 391: 219–234, 1987.

Alger BE, Nicoll RA: Pharmacological evidence for two kinds of GABA receptor on rat hippocampal pyramidal cells studied *in vitro*. J Physiol (Lond) 328: 125–141, 1982.

Allen GI, Eccles JC, Nicoll RA, Oshima T, Rubia FJ: The ionic mechanisms concerned in generating the i.p.s.p.s. of hippocampal pyramidal cells. Proc R Soc Lond [Biol] 198: 363–384, 1977.

Alvarez-Leefmans FJ: Intracellular Cl^- regulation and synaptic inhibition in vertebrate and invertebrate neurons. In Alvarez-Leefmans FJ, Russell JM (eds): "Chloride channels and Carriers in Nerve, Muscle, and Glial Cells." New York: Plenum Press, 1990, pp. 109–158.

Alvarez-Leefmans FJ, Gamino SM, Giraldez F, Nogueron I: Intracellular chloride regulation in amphibian dorsal root ganglion neurones studied with ion-selective microelectrodes. J Physiol (Lond) 406: 225–246, 1988.

Andersen P, Dingledine R, Gjerstad L, Langmoen IA, Laursen AM: Two different responses of hippocampal pyramidal cells to application of gamma-aminobutyric acid. J Physiol (Lond) 305: 279–296, 1980.

Andrade R, Malenka RC, Nicoll RA: A G protein couples serotonin and GABA-receptors to the same channels in hippocampus. Science 234: 1261–1265, 1988.

Araki T, Ito M, Oscarsson O: Anion permeability of the synaptic and non-synaptic motoneurone membrane. J Physiol (Lond) 159: 410–435, 1961.

Asano T, Ui M, Ogasawara N: Prevention of the agonist binding to γ-aminobutyric acid B receptors by guanine nucleotides and islet-activating protein, pertussis toxin, in bovine cerebral cortex. J Biol Chem 260: 12653–12658, 1985.

Avoli M, Perreault P: A GABAergic depolarizing potential in the hippocampus disclosed by the convulsant 4-aminopyridine. Brain Res 400: 191–195, 1987.

Awapara J, Landua AJ, Fuerst R, Seale B: Free γ-aminobutyric acid in brain. J Biol Chem 187: 35–39, 1950.

Bazemore AW, Elliott KAC, Florey E: Isolation of factor I. J. Neurochem 1: 334–339, 1957.

Ben-Ari Y, Krnjević K, Reinhardt W: Hippocampal seizures and failure of inhibition. Can J Physiol Pharmacol 57: 1462–1466, 1979.

Ben-Ari Y, Krnjević K, Reiffenstein RJ, Reinhardt W: Inhibitory conductance changes and action of γ-aminobutyrate in rat hippocampus. Neuroscience 6: 2445–2463, 1981a.

Ben-Ari Y, Krnjević K, Reinhardt W, Ropert N: Intracellular observations on the disinhibitory action of acetylcholine in the hippocampus. Neuroscience 6: 2475–2484, 1981b.

Ben-Ari Y, Cherubini E: Brief anoxic episodes induce long-lasting changes in synaptic properties of rat CA3 hippocampal neurons. Neurosci Lett 90: 273–278, 1988.

Ben-Ari Y, Cherubini E, Coradetti R, Gaiarsa JL: Giant synaptic potentials in immature rat CA3 hippocampal neurones. J Physiol (Lond) 416: 303–325, 1989.

Benveniste H, Drejer J, Schousboe A, Diemer NH: Elevation of the extracellular concentrations of glutamate and aspartate in rat hippocampus during transient cerebral ischemia monitored by intracerebral microdialysis. J Neurochem 43: 1369–1374, 1984.

Blair LAC, Levitan ES, Marshall J, Dionne VE, Barnard EA: Single subunits of the GABA$_A$ receptor form ion channels with properties of the native receptor. Science 242: 577–579, 1988.

Bormann J: Electrophysiology of GABA-A and GABA-B receptor subtypes. Trends Neurosci 11: 112–116, 1988.

Bormann J, Hamill OP, Sakmann B: Mechanism of anion permeation through channels gated by glycine and gamma-aminobutyric acid in mouse cultured spinal motoneurones. J Physiol (Lond) 385: 243–286, 1987.

Bowery NG, Price GW, Hudson AL, Hill DR, Wilkin GP, Turnbull MJ: GABA receptor multiplicity. Visualization of different receptor types in the mammalian CNS. Neuropharmacology 23: 219–231, 1984.

Bowery NG, Hill DR, Moratalla R: Neurochemistry and autoradiography of GABA$_B$ receptors in mammalian brain: Second-messenger system(s). In Barnard EA, Costa E (eds): "Neurochemistry and Autoradiography of GABA$_B$ Receptors in Mammalian Brain: Second-messenger system(s)." New York: Raven Press, 1989, pp. 159–172.

Bowery NG, Hill DR, Hudson AL, Doble A, Middlemiss DN, Shaw J, Turnbull M: (-) Baclofen decreases neurotransmitter release in the mammalian CNS by an action at a novel GABA receptor. Nature 283: 92–94, 1980.

Casullo J, Krnjević K: Glial potentials in hippocampus. Can J Physiol Pharmacol 65: 847–855, 1987.

Chavez-Noriega LE, Bliss TVP, Halliwell JV: The EPSP-spike (E-S) component of long-term potentiation in the rat hippocampal slice is modulated by GABAergic but not cholinergic mechanisms. Neurosci Lett 104: 58–64, 1989.

Choi DW, Farb DH, Fischbach GD: Chlordiazepoxide selectively augments GABA action in spinal cord cell cultures. Nature 269: 342–344, 1977.

Cockcroft S, Stutchfield J: G-proteins, the inositol lipid signalling pathway, and secretion. Philos Trans R Soc Lond [Biol] B 320: 247–265, 1988.

Collingridge GL, Herron CE, Lester RAJ: Frequency-dependent-N-methyl-D-aspartate receptor-mediated synaptic transmission in rat hippocampus. J Physiol (Lond) 399: 301–312, 1988.

Connor JA, Wadman WJ, Hockberger PE, Wong RKS: Sustained dendritic gradients of Ca^{2+} induced by excitatory amino acids in CA1 hippocampal neurons. Science 240: 649–653, 1988.

Connors BW, Malenka RC, Silva LR: Two inhibitory postsynaptic potentials, and GABA-A and GABA-B receptor-mediated responses in neocortex of rat and cat. J Physiol (Lond) 406: 443–468, 1988.

Coombs JS, Eccles JC, Fatt P: The specific ionic conductances and the ionic movements across the motoneuronal membrane that produce the inhibitory post-synaptic potential. J Physiol (Lond) 130: 326–373, 1955.

Cottrell GA, Lambert JJ, Peters JA: Modulation of GABA-$_A$ receptor activity by alphaxalone. B J Pharmacol 90: 491–500, 1987.

Curtis DR, Duggan AW, Felix D, Johnston GAR, McLennan H: Antagonism between bicuculline and GABA in the cat brain. Brain Res 33: 57–73, 1971.

Curtis DR, Phillis JW, Watkins JC: The depression of spinal neurones by γ-amino-n-butyric acid and β-alanine. J Physiol (Lond) 146: 185–203, 1959.

Curtis DR, Johnston GAR: Aminoacid transmitters in the mammalian central nervous system. Ergeb Physiol 69: 97–188, 1974.

Dalkara T: Nipecotic acid, an uptake blocker, prevents the fading of the γ-aminobutyric acid effect. Brain Res 366: 314–319, 1986.

Davidoff RA, Sears ES: The effects of lioresal on synaptic activity in the isolated spinal cord. Neurology 24: 957–963, 1974.

Davies CH, Davies SN, Collingridge GL: Paired-pulse depression of monosynaptic GABA-mediated inhibitory postsynaptic responses in rat hippocampus. J Physiol 424: 513–531, 1990.

Davies CH, Starkey SJ, Pozza MF, Collingride GL: GABA autoreceptors regulate the induction of LTP. Nature 349:609–611, 1991.

DeFeudis FV: Muscimol and GABA-receptors: Basic studies and therapeutic implications. Rev Pure Applied Pharmacol Sci. 3: 319–379, 1982.

DeFeudis FV: GABA$_A$ receptors of the central nervous system: An update. Drugs of Today 25: 525–540, 1989.

Deisz RA, Prince DA: Frequency-dependent depression of inhibition in guinea-pig neocortex *in vitro* by GABA-B receptor feedback on GABA release. J Physiol (Lond) 412: 513–541, 1989.

Dingledine R, Hynes MA, King GL: Involvement of N-methyl-D-aspartate receptors in epileptiform bursting in the rat hippocampal slice. J Physiol (Lond) 380: 175–189, 1986.

Dingledine R, Boland LM, Chamberlin NL, Kawasaki K, Kleckner NW, Traynelis SF, Versoorn TA: Amino acid receptors and uptake systems in the mammalian central nervous system. Crit Rev Neurobiol 4: 1–96, 1988.

Dolphin AC, Scott RH: Calcium channel currents and their inhibition by (-)-baclofen in rat sensory neurones: modulation by guanine nucleotides. J Physiol (Lond) 386: 1–17, 1987.

Dreifuss JJ. Kelly JS, Krnjević K: Cortical inhibition and γ-aminobutyric acid. Exp Brain Res 9: 137–154, 1969.

Dunlap K, Fischbach GD: Neurotransmitters decrease the calcium component of sensory neurone action potentials. Nature 276: 837–839, 1978.

Dutar P, Nicoll RA: Pre- and postsynaptic GABA$_B$- receptors in the hippocampus have different pharmacological properties. Neuron 1: 585–591, 1988a.

Dutar P, Nicoll RA: A physiological role of GABA$_B$- receptors in the central nervous system. Nature 332: 156–158, 1988b.

Dykes RW, Landry P, Metherate R, Hicks TP: Functional role of GABA in cat primary somatosensory cortex: shaping receptive fields of cortical neurons. J Neurophysiol 52: 1066–1093, 1984.

Ebert U, Krnjević K: Systemic CI-966, a new GABA uptake blocker, enhances GABA-action in CA1 pyramidal layer *in situ*. Can J Physiol Pharmacol 68: 1194–1199, 1990.

Eccles JC: "The Physiology of Synapses." Berlin; Springer Verlag. 1964, pp. 316.

Edwards FR, Harrison PJ, Jack JJB, Kullmann DM: Reduction by baclofen of monosynaptic EPSPs in lumbosacral motoneurones of the anaesthetized cat. J Physiol (Lond) 416: 539–556, 1989.

Elliott KAC, van Gelder NM: Occlusion and metabolism of γ-aminobutyric acid of brain tissue. J Neurochem 3: 28–40, 1958.

Enna SJ, Gallagher JP: Biochemical and electrophysiological characteristics of mammalian GABA receptors. Intern Rev Neurobiol 24:181–212, 1983.

Erecińska M: The neurotransmitter amino acid transport systems. A fresh outlook on an old problem. Biochem Pharmacol 36: 3547–3555, 1987.

Fox S, Krnjević K, Morris ME, Puil E, Werman R: Action of baclofen on mammalian synaptic transmission. Neuroscience 3: 495–515, 1978.

Fujiwara N, Higashi H, Shimoji K, Yoshimura M: Effects of hypoxia on rat hippocampal neurones *in vitro*. J Physiol (Lond) 384: 131–151, 1987.

Gähwiler BH, Brown DA: GABA—receptor-activated K^+ current in voltage-clamped CA3 pyramidal cells in hippocampal cultures. Proc Natl Acad Sci USA 82: 1558–1562, 1985.

Gale K: Mechanisms of seizure control mediated by γ-aminobutyric acid: role of the substantia nigra. Fedn Proc 44: 2414–2424, 1985.

Giaretta D, Avoli M, Gloor P: Intracellular recordings in pericruciate neurons during spike and wave discharges of feline generalized penicillin epilepsy. Brain Res 405: 68–79, 1987.

Godfraind JM, Krnjević K, Pumain R: Doubtful value of bicuculline as a specific antagonist of GABA. Nature 228: 675–676, 1970.

Gray R, Johnston D: Rectification of single GABA-gated chloride channels in adult hippocampal neurons. J Neurophysiol 54: 134–142, 1985.

Griffith WH, Brown TH, Johnston D: Voltage-clamp analysis of synaptic inhibition during long-term potentiation in hippocampus. J Neurophysiol 44: 767–775, 1986.

Haas HL: Cholinergic disinhibition in hippocampal slices of the rat. Brain Res 233: 200–204, 1982.

Haas HL, Rose G: The role of inhibitory mechanisms in hippocampal long-term potentiation. Neurosci Lett 47: 301–306, 1984.

Haefely W: Pharmacology of the allosteric modulation of GABA$_A$ receptors by benzodiazepine receptor ligands. In Barnard EA, Costa E (eds): "Allosteric Modulation of Amino Acid Receptors: Therapeutic Implications." Vol. 1, New York: Raven Press, 1989, pp 47–69.

Haefely W, Pieri L, Polc P, Schaffner R: General Pharmacology and Neuropharmacology of Benzodiazepine Derivatives. In Hoffmeister F, Stille G (eds): "Handbook of Experimental Pharmacology." Vol. 55 II, 1981, pp. 13–262, Heidelberg: Springer-Verlag.

Hamill OP, Bormann J, Sakmann B: Activation of multiple-conductance state chloride channels in spinal neurones by glycine and GABA. Nature 305: 805–808, 1983.

Harrison NL, Lange GD, Barker JL: (-) Baclofen activates presynaptic GABA-B receptors on GABAergic inhibitory neurons from embryonic rat hippocampus. Neurosci Lett 85: 105–109, 1988.

Henn FA, Hamburger A: Glial cell function: uptake of transmitter substances. Proc Nat Acad Sci USA 68: 2686–2690, 1971.

Inoue M, Matsuo T, Ogata N: Characterization of pre- and postsynaptic actions of (-)-baclofen in the guinea-pig hippocampus *in vitro*. Br J Pharmacol 84: 843–851, 1985a.

Inoue M, Matsuo T, Ogata N: Baclofen activates voltage-dependent and 4-aminopyridine sensitive K^+ conductance in guinea-pig hippocampal pyramidal cells maintained *in vitro*. Br J Pharmacol 84: 833–841, 1985b.

Inoue M, Oomura Y, Yakushi T, Akaike N: Intracellular calcium ions decrease the affinity of the GABA receptor. Nature 324: 156–158, 1986.

Ito M: "The Cerebellum and Neural Control." New York: Raven Press, 1984, pp. 580.

Ito M, Yoshida M: The origin of cerebellar-induced inhibition of Deiters' neurones. I. Monosynaptic initiation of the inhibitory postsynaptic potentials. Exp Brain Res 2: 330–349, 1966.

Johnston GAR, Stephanson AL, Twitchin B: Uptake and release of nipecotic acid by rat brain. J Neurochem 26: 83–87, 1976.

Kaila K, Pasternack M, Saarikoski J, Voipio J: Influence of GABA-gated bicarbonate conductance on potential, current and intracellular chloride in crayfish muscle fibres. J Physiol (Lond) 416: 161–181, 1989.

Kandel ER, Spencer WA, Brinley FJ: Electrophysiology of hippocampal neurons. I. Sequential invasion and synaptic organization. J Neurophysiol 24: 225–242, 1961.

Kanner BI, Radian R: Ion-coupled neurotransmitter transport across the synaptic plasma membrane. Ann NY Acad Sci 456: 153–161, 1985.

Kapur J, Stringer JL, Lothman EW: Evidence that repetitive seizures in the hippocampus cause a lasting reduction of GABAergic inhibition. J Neurophysiol 61: 417–426, 1989.

Karbon EW, Enna SJ: Characterization of the relationship between γ-aminobutyric acid B agonists and transmitter-coupled cyclic nucleotide-generating systems in rat brain. Mol Pharmacol 27: 53–59, 1985.

Katz B, Thesleff S: A study of the "desensitization" produced by acetylcholine at the motor end-plate. J Physiol (Lond) 138: 63–80, 1957.

Kelly JS, Krnjević K, Morris ME, Yim GKW: Anionic permeability of cortical neurones. Exp. Brain Res 7: 11–31, 1969.

Kerr DIB, Ong J, Johnston GAR, Abbenante J, Prager RH: 2-Hydroxysaclofen: an improved antagonist at central and peripheral GABA$_B$ receptors. Neurosci Lett 92: 92–96, 1988.

Kerr DIB, Ong J, Prager RH, Gynther BD, Curtis DR: Phaclofen: a peripheral and central baclofen antagonist. Brain Res: 405: 150–154, 1987.

Kettenmann H, Backus KH, Schachner M: γ-aminobutyric acid opens Cl-channels in cultured astrocytes. Brain Res 404: 1–9, 1987.

Korn SJ, Dingledine R: Inhibition of GABA uptake in the rat hippocampal slice. Brain Res 368: 247–255, 1986.

Krnjević K: Chemical nature of synaptic transmission in vertebrates. Physiol Rev 54: 418–540, 1972.

Krnjević K: Inhibitory action of GABA and GABA mimetics on vertebrate neurones. In Roberts E, Chase TN, Tower DB (eds): "GABA in Nervous System Function." New York: Raven Press, 1976, pp. 269–281.

Krnjević K: GABA-mediated inhibitory mechanisms in relation to epileptic discharges. In Jasper HH, van Gelder NM (eds): "Basic Mechanisms of Neuronal Hyperexcitability." New York: Alan R. Liss, Inc., 1983, pp. 249–280.

Krnjević K: Some functional consequences of GABA uptake by brain cells. Neurosci Lett 47: 283–287, 1984.

Krnjević K, Miledi R: Presynaptic failure of neuromuscular propagation in rats. J Physiol (Lond) 149: 1–22, 1959.

Krnjević K, Schwartz S: The action of gamma-aminobutyric acid on cortical neurones. Exp Brain Res 3: 320–336, 1967.

Krnjević K, Morris ME, Reiffenstein RJ: Changes in extracellular Ca^{2+} and K^+ activity accompanying hippocampal discharges. Can J Physiol Pharmacol 58: 579–583, 1980.

Krnjević K, Morris ME, Ropert N: Changes in free calcium ion concentration recorded inside hippocampal pyramidal cells *in situ*. Brain Res 374: 1–11, 1986.

Krnjević K, Xu YZ, Zhang L: Anoxic block of GABAergic IPSPs. Neurochem Res 16: 279–283, 1991.

Krogsgaard-Larsen P: Inhibition of the GABA uptake systems. Mol Cell Biochem 31: 105–121, 1980.

Lambert NA, Harrison NL, Kerr DIB, Ong J, Prager RH, Teyler TJ: Blockade of the later IPSP in rat CA1 hippocampal neurons by 2-hydroxy-saclofen. Neurosci Lett 107: 125–128, 1989.

Lebeda FJ, Hablitz JJ, Johnston D: Antagonism of GABA-mediated responses by d-tubocurarine in hippocampal neurons. J Neurophysiol 48: 622–632, 1982.

Leblond J, Krnjević K: Hypoxic changes in hippocampal neurons. J Neurophysiol 62: 1–14, 1989.

Lipton P, Whittingham TS: Energy metabolism and brain slice function. In Dingledine R (ed): "Brain Slices." New York: Plenum Press, 1984, pp. 113–153.

Lux HD, Loracher C, Neher E: The action of ammonium on postsynaptic inhibition of cat spinal motoneurons. Exp Brain Res 11: 431–447, 1970.

Macdonald RL, Barker JF: Pentylenetetrazol and penicillin are selective antagonists of GABA-mediated post-synaptic inhibition in cultured mammalian neurones. Nature 267: 720–721, 1977.

Macdonald RL, McLean MJ, Skerritt JH: Anticonvulsant drug mechanisms of action. Fed Proc 44: 2634–2639, 1985.

Macdonald RL, Skerritt JH, Werz MA: Barbiturate and benzodiazepine actions on mouse neurons in cell culture. In Roth SH, Miller KW (eds): "Molecular and Cellular Mechanisms of Anesthetics." New York: Plenum Press, 1986, pp. 17–25.

MacVicar BA, Tse FWY, Crichton SA, Kettenmann H: GABA-activated Cl⁻ channels in astrocytes of hippocampal slices. J Neurosci 9: 3577–3583, 1989.

Majewska MD, Harrison NL, Schwartz RD, Barker JL, Paul SM: Steroid hormone metabolites are barbiturate-like modulators of the GABA receptor. Science 232: 1004–1007, 1986.

Marciani MG, Louvel J, Heinemann U: Aspartate-induced changes in extracellular free calcium in "in vitro" hippocampal slices of rats. Brain Res 238: 272–277, 1982.

Martin IL: The benzodiazepines and their receptors: 25 years of progress. Neuropharmacol 26: 957–970, 1987.

Massey SC, Redburn DA: Transmitter circuits in the vertebrate retina. Prog Neurobiol 28: 55–96, 1987.

McCarren M, Alger BE: Use-dependent depression of IPSPs in rat hippocampal pyramidal cells in vitro. J Neurophysiol 53: 557–571, 1985.

McCormick DA: GABA as an inhibitory neurotransmitter in human cerebral cortex. J Neurophysiol 62: 1018–1027, 1989.

McGeer PL, Eccles JC, McGeer EG: "Molecular Neurobiology of the Mammalian Brain." New York: Plenum Press, pp. 197–234, 1987.

McLaughlin BJ, Wood JG, Saito K, Barber R, Vaughn JE, Roberts E, Wu J-Y: The fine structural localization of glutamate decarboxylase in synaptic terminals of rodent cerebellum. Brain Res 76: 377–391, 1974.

McNamara JO, Galloway MT, Rigsbee LC, Shin C: Evidence implicating substantia nigra in regulation of kindled seizure threshold. J Neurosci 4: 2410–2417, 1984.

Miledi R, Parker I, Sumikawa K: Synthesis of chick brain GABA receptors by frog oocytes. Proc R Soc Lond [Biol] B216: 509–515, 1982.

Misgeld U, Deisz RA, Dodt HU, Lux HD: The role of chloride transport in post-synaptic inhibition of hippocampal neurons. Science 232: 1413–1415, 1986.

Misgeld U, Klee MR, Zeise ML: Differences in burst characteristics and drug sensitivity between CA3 neurons and granule cells. In Klee MR, Lux HD,

Speckmann E-J (eds): "Physiology and Pharmacology of Epileptogenic Phenomena." New York: Raven Press, 1982, pp. 131–139.

Möhler H, Okada T: Benzodiazepine receptor: Demonstration in the central nervous system. Science 198: 849–851, 1977.

Monod J, Wyman J, Changeux J-P: On the nature of allosteric transitions: a plausible model. J Mol Biol 12: 88–118, 1965.

Morris ME, Di Constanzo GA, Barolet A, Sheridan PJ: Role of K$^+$ in GABA (γ-aminobutyric acid)-evoked depolarization of peripheral nerve. Brain Res 278: 127–135, 1983.

Mulle C, Madariaga A, Deschênes M: Morphology and electrophysiological properties of reticularis thalami neurons in cat: *In vivo* study of a thalamic pacemaker. J Neurosci 6: 2134–2145, 1990.

Muller D, Joly M, Lynch G: Contributions of quisqualate and NMDA receptors to the induction and expression of LTP. Science 242: 1694–1697, 1988.

Newberry NR, Nicoll RA: Comparison of the action of baclofen with γ-aminobutyric acid on rat hippocampal pyramidal cells. J Physiol 360: 161–185, 1985.

Nicholls DG: Release of glutamate, aspartate, and γ-aminobutyric acid from isolated nerve terminals. J Neurochem 52: 331–341, 1989.

Nicoll RA: The effects of anaesthetics on synaptic excitation and inhibition in the olfactory bulb. J Physiol 223: 803–814, 1972.

Nicoll RA, Madison DV: General anaesthetics hyperpolarize neurons in the vertebrate central nervous system. Science 217: 1055–1057, 1982.

Nistri A, Constanti A: Pharmacological characterization of different types of GABA and glutamate receptors in vertebrates and invertebrates. Prog Neurobiol 13: 117–235, 1979.

Numann RE, Wong RKS: Voltage-clamp study of GABA response desensitization in single pyramidal cells dissociated from the hippocampus of adult guinea pigs. Neurosci Lett 47: 289–294, 1984.

Olpe H-R, Baudry M, Fagni L, Lynch G: The blocking action of baclofen on excitatory transmission in the rat hippocampal slice. J Neurosci 2: 698–703, 1982.

Olsen RW: Drug interactions at the GABA receptor-ionophore complex. Annu Rev Pharmacol Toxicol 22: 247–277, 1982.

Olsen RW, Wong EHF, Stauber GB, King RG: Biochemical pharmacology of the γ-aminobutyric acid receptor/ionophore protein. Fed Proc 43: 2773–2778, 1984.

Osmanović SS, Shefner SA: Baclofen increases the potassium conductance of rat locus coeruleus neurons recorded in brain slices. Brain Res 438: 124–136, 1988.

Owen DG, Barker, JL, Segal M, Study RE: Postsynaptic actions of pentobarbital in cultured mouse spinal neurons and rat hippocampal neurons. In Roth SH, Miller KW (eds): "Molecular and Cellular Mechanisms of Anesthetics." New York: Plenum Press, 1986, pp. 27–41.

Pavlov IP: "Conditioned Reflexes." Oxford: University Press, 1927.

Peet MJ, McLennan H: Pre- and postsynaptic actions of baclofen: blockade of the late synaptically-evoked hyperpolarization of CA1 hippocampal neurones. Exp Brain Res 61: 567–574, 1986.

Peng Y, Frank E: Activation of GABA-B receptors causes presynaptic inhibition at synapses between muscle spindle afferents and motoneurons in the spinal cord of bullfrogs. J Neurosci 9: 1502–1515, 1989.

Phillis JW, Wu PH: Interactions between the benzodiazepines, methylxanthines and adenosine. Can J Neurol Sci 7: 247–249, 1980.

Pierau F-K, Zimmermann P: Action of GABA-derivative on post-synaptic potentials and membrane properties of cat's spinal motoneurones. Brain Res 54: 376–380, 1973.

Polc P, Bonetti EP, Schaffner R, Haefely W: A three-state model of the benzodiazepine receptor explains the interaction between the benzodiazepine antagonist Ro-15–1788, benzodiazepine tranquilizers, β-carbolines, and phenobarbitone. Naunyn Schmiedebergs Arch Pharmacol 321: 260–264, 1982.

Polc P: Electrophysiology of benzodiazepine receptor ligands: Multiple mechanisms and sites of action. Prog Neurobiol 31: 349–423, 1988.

Potashner SJ: Baclofen: effects on amino acid release and metabolism in slices of guinea pig cerebral cortex. J Neurochem 32: 103–109, 1979.

Potashner SJ, Gerard G: Kainate-enhanced release of D-[^{3}H] aspartate from cerebral cortex and striatum: reversal by baclofen and pentobarbital. J Neurochem 40: 1548–1557, 1983.

Pritchett DB, Sontheimer H, Gorman CM, Kettenmann H, Seeburg PH, Schofield PR: Transient expression shows ligand gating and allosteric potentiation of GABA$_A$ receptor subunits. Science 242: 1306–1308, 1988.

Pulsinelli WA: Selective neuronal vulnerability: morphological and molecular characteristics. Prog Brain Res 63: 29–37, 1985.

Raabe W, Gumnit RJ: Disinhibition in cat motor cortex by ammonia. J Neurophysiol 38: 347–355, 1975.

Rekling JC, Jahnse H, Laursen AM: The effect of two lipophilic γ-aminobutyric acid uptake blockers in CA1 of the rat hippocampal slice. Br J Pharmacol 99: 103–106, 1990.

Ribak CE: GABAergic abnormalities occur in experimental models of focal and genetic epilepsy. J Mind Behavior 8: 605–618, 1987.

Roberts E: Disinhibition as an organizing principle in the nervous system—The role of the GABA system. Application to neurological and psychiatric disorders. In Robert E, Chase TN, Tower DB (eds): "GABA in Nervous System Function." New York: Raven Press, 1976, pp. 515–539.

Roberts E, Frankel S: γ-Aminobutyric acid in brain: its formation from glutamic acid. J Biol Chem 187: 55–63, 1950.

Rodbell M: The role of hormone receptors and GTP-regulatory proteins in membrane transduction. Nature 284: 17–22, 1980.

Rovira C, Ben-Ari Y, Cherubini E: Somatic and dendritic actions of γ-aminobutyric acid agonists and uptake blockers in the hippocampus *in vivo*. Neuroscience 12: 543–555, 1984.

Ryan AF, Schwartz IR: Nipecotic acid: preferential accumulation in the cochlea by GABA uptake systems and selective retrograde transport to brainstem. Brain Res 399: 399–403, 1986.

Sato TN, Neale JH: The type I and type II γ-aminobutyric acid/benzodiazepine receptor: 1. Purification and two-dimensional electrophoretic analysis of the receptor from cortex and cerebellum. Biochem Biophys Res Comm 146: 568–574, 1987.

Sawynok J: Baclofen activates two distinct receptors in the rat spinal cord and guinea pig ileum. Neuropharmacology 25: 795–798, 1986.

Schofield PR, Darlison MG, Fujita N, Burt DR, Stephenson FA, Rodriguez H, Rhee LM, Ramachandran J, Reale V, Glencorse TA, Seeburg PH, Barnard EA: Sequence and functional expression of the GABA$_A$ receptor shows a ligand-gated receptor super-family. Nature 328: 221–227, 1987.

Scholfield CN: Potentiation of inhibition by general anaesthetics in neurones of the olfactory cortex *in vitro*. Pflugers Arch. 383: 249–255, 1980.

Schultz, SG, Curran PF: Coupled transport of sodium and organic solutes. Physiol Rev 50: 637–718, 1970.

Schwartz, EA: Depolarization without calcium can release γ-aminobutyric acid from a retinal neuron. Science 238: 350–355, 1987.

Schwartz-Giblin S, Canonaco M, McEwen BS, Pfaff DW: Effects of *in vivo* estradiol and progesterone on tritiated flunitrazepam binding in rat spinal cord. Neuroscience 25: 249–257, 1988.

Segal M: Rat hippocampal neurons in culture: responses to electrical and chemical stimuli. J Neurophysiol 50: 1249–1264, 1983.

Segal M, Barker JL: Rat hippocampal neurons in culture: Voltage-clamp analysis of inhibitory synaptic connections. J Neurophysiol 52: 469–487, 1984.

Selye H: Anesthetic effect of steroid hormones. Proc Soc Exp Biol Med 46: 116–121, 1941.

Shiroya T, Fukunaga R, Akashi K, Shimada N, Takagi Y, Nishino T, Hara M, Inagaki C: An ATP-driven Cl⁻ pump in the brain. J Biol Chem 264: 17416–17421, 1989.

Sillito AM: Functional considerations of the operation of GABAergic inhibitory processes in the visual cortex. In Jones EG, Peters A (eds): "Cerebral Cortex." New York: Plenum Press, 1984, pp. 91–117.

Snow RW, Taylor CP, Dudek FE: Electrophysiological and optical changes in slices of rat hippocampus during spreading depression. J Neurophysiol 50: 561–572, 1983.

Somjen GG: Extracellular potassium in the mammalian central nervous system. Annu Rev Physiol 41: 159–177, 1979.

Squires RF, Benson DI, Braestrup C, Coupet J, Klepner CA, Meyers V, Beer B: Some properties of brain specific benzodiazepine receptors: New evidence for multiple receptors. Pharmacol Biochem Behav 10: 825–830, 1979.

Stelzer A, Kay AR, Wong RKS: GABA-A receptor function in hippocampal cells is maintained by phosphorylation factors. Science 241: 339–341, 1988.

Stelzer A, Slater NT, ten Bruggencate G: Activation of NMDA receptors blocks GABAergic inhibition in an in vitro model of epilepsy. Nature 326: 698–701, 1987.

Steriade M, Llinas R: The functional states of the thalamus and the associated neuronal interplay. Physiol Revs. 68: 649–742, 1988.

Stewart PA: "How to Understand Acid-Base. A Quantitative Acid-Base Primer for Biology and Medicine." New York: Elsevier, 1981, pp. 186.

Sugiyama H, Ito I, Hirono C: A new type of glutamate receptor linked to inositol phospholipid metabolism. Nature 325: 531–533, 1987.

Sumikawa K, Parker I, Miledi R: Partial purification and functional expression of brain mRNAs coding for neurotransmitter receptors and voltage-operated channels. Proc Natl Acad Sci (USA) 81: 7994–7998, 1984.

Swadlow HA, Kocsis JD, Waxman SG: Modulation of impulse conduction along the axonal tree. Annu Rev Biophys Bioeng 9: 143–179, 1980.

Taylor CP, Vartanian MG: Anticonvulsant and behavioral effects of CI-966 and PD 126141, potent GABA-uptake inhibitors, in animals. Soc Neurosci Abstr 15: 667, 1989.

Thalmann RH: Evidence that guanosine triphosphate (GTP)-binding proteins control a synaptic response in brain: effect of pertussis toxin and GTPγS on the late inhibitory post-synaptic potential of hippocampal CA3 neurons. J Neurosci 8: 4589–4602, 1988.

Thompson SM, Gähwiler BH: Activity-dependent disinhibition. II. Effects of extracellular potassium, furosemide, and membrane potential on E_{cl}⁻ in hippocampal CA3 neurons. J Neurophysiol 61: 512–521, 1989a.

Thompson SM, Gähwiler BH: Activity-dependent disinhibition. III. Desensitization and GABA receptor-mediated presynaptic inhibition in the hippocampus *in vitro*. J Neurophysiol 61: 524–533, 1989b.

Twyman RE, Rogers CJ, Macdonald RI: Differential regulation of γ-aminobutyric acid receptor channels by diazepam and phenobarbital. Ann Neurol 25: 213–220, 1989.

Weiss DS, Barnes EM, Hablitz JJ: Whole-cell and single-channel recordings of GABA-gated currents in cultured chick cerebral neurons. J Neurophysiol 59: 495–513, 1988.

Wong RKS, Watkins DJ: Cellular factors influencing GABA response in hippocampal pyramidal cells. J Neurophysiol 48: 938–951, 1982.

Yoshida M, Obata K: Actions of putative neurotransmitters on cat pallidal neurons. In Ryall RW and Kelly JS (eds): "Iontophoresis and Transmitter Mechanisms in the Mammalian Central Nervous System." Amsterdam: Elsevier/North-Holland Biomedical Press, 1978, pp. 71–74.

GABA Mechanisms in Epilepsy, pages 89–104
© 1991 Wiley-Liss, Inc.

4
Antiepileptic Drug Regulation of GABA$_A$ Receptor Channels

ROY E. TWYMAN AND ROBERT L. MACDONALD

Departments of Neurology (R.E.T., R.L.M.) and Physiology (R.L.M.), University of Michigan Medical Center, Ann Arbor, MI 48104

INTRODUCTION

All antiepileptic drugs (AEDs) currently in use alter neuronal excitability. Most affect membrane associated ion channels that selectively allow passage of sodium, calcium, or chloride ions (Courtney and Etter, 1983; Macdonald and McLean, 1986; Macdonald, 1989; Macdonald and Meldrum, 1989). Several clinically different types of epilepsy, such as generalized tonic-clonic, generalized absence, and partial and myoclonic seizures exist, and several classes of AEDs have been used in their treatment (Macdonald, 1989). These classes include hydantoins, iminostilbenes, barbiturates, fatty acids, benzodiazepines, and succinimides. An association exists between the class of AED and seizure type. Hydantoins (phenytoin) and iminostilbenes (carbamazepine) are used primarily in the treatment of generalized tonic-clonic and partial seizures. A succinimide (ethosuximide) is used exclusively in the treatment of absence seizures. Barbiturates (phenobarbital and primidone), benzodiazepines (diazepam, lorazepam, clonazepam, nitrazepam), and a fatty acid (sodium valproate) can be used against a variety of seizure types.

In concordance with treatment selectivity, the mechanisms of action of the different classes of AEDs also differ (Macdonald and McLean, 1986). Abnormal neuronal excitability appears to be the basis for the development of seizures (see Crill, this volume), and AEDs have been demonstrated to differentially modulate neuronal excitability. The sensitivity of neuronal electrical activity can be modulated by decreasing excitation or increasing inhibition. One indication of neuronal excitability is sustained, high-frequency, repetitive firing of action potentials (SRF). At supratherapeutic concentrations, commonly used AEDs effective in the treatment of generalized tonic-clonic or partial seizures (phenytoin, carbamazepine, valproate, phenobarbital, primidone, and benzodiazepines) reduce SRF (McLean and Macdonald, 1986; Macdonald and McLean, 1986). On the other hand, ethosuximide does not reduce SRF (McLean and Macdonald, 1986). Only phenytoin, carbamazepine, and valproate reduce SRF at concentrations that approximate therapeutic free-serum concentrations. The effect of phenytoin, carbamazepine, and valproate on SRF has been related to their ability to regulate sodium channel function (Macdonald, 1989). Phenytoin and carbamazepine have been demonstrated to bind to a batrachotoxin (BTX-B) binding site related to the activation of sodium channels (Willow et al., 1984). Valproate, however, does not bind to the BTX-B site, and therefore, its binding site on the sodium channel remains unknown (Willow and Catterall, 1982).

Phenytoin, the barbiturates and the benzodiazepine have also been shown to regulate voltage-dependent calcium channels. Calcium current influx through these channels appears to be important in the regulation of presynaptic release of neurotransmitter. Three types of calcium currents exist and reflect the existence of possibly three types of calcium channels (N, L, and T) (Nowycky et al., 1985). At anesthetic plasma concentrations of pentobarbital and phenobarbital, calcium currents of the N- and L-types are reduced, but T-type currents are not (Gross and Macdonald, 1987; 1988). T-currents might be important pacemaker currents in thalamic neurons and may underlie the 3-Hz spike and wave rhythm seen in electroencephalograms of patients with generalized absence seizures (Coulter et al., 1989a). Ethosuximide, trimethiadione metabolite dimethadione, and sodium valproate have been demonstrated to decrease T-currents (Coulter et al., 1989b; Kelly et al., 1990). It has been proposed that reduction of T-current disrupts the rhythmic firing of thalamic neurons and also the neuronal origin of the spike and wave discharge (Coulter et al., 1989a,b). Benzodiazepines are also used in the treatment of generalized absence seizures but do not appear to alter T-currents (Gross et al., 1989). The reason for the difference between these agents and ethosuximide is unknown. Other calcium channel blockers including nimodipine and flunarizine (Overweg, et al., 1984; Meyer et al., 1986) are currently under clinical investigation for the control of seizures. It is not certain that these drugs will be useful clinically.

Excitatory synaptic function is mediated primarily through the excitatory neurotransmitters, glutamate and possibly aspartate, which can activate at least two different glutamate receptor subtypes, N-methyl-D-aspartate (NMDA) and non-NMDA receptors (Mayer and Westbrook, 1987). Recently, NMDA receptors have been proposed to be important in the development and regulation of seizures. NMDA receptors are associated with a channel permeable to sodium and calcium ions and can be blocked in a voltage-dependent fashion by magnesium ions. Selective blockade of NMDA receptors by compounds such as MK-801 and 2-amino-5-phosphovalerate (APV) have been shown to be potent anticonvulsants in animal models of epilepsy (Croucher et al., 1982; Meldrum et al., 1983). Clinical usefulness of agents that modulate NMDA receptors remains to be proven.

A decrease in neuronal excitability can be achieved by increasing neuronal inhibition at the post-synaptic membrane. In supraspinal neurons, this inhibition is mediated by GABA activation of GABA$_A$ receptor chloride channels and GABA receptor potassium or alcium channels. Only the GABA$_A$ receptor will be discussed here. Several approaches have been taken to regulate GABA receptor function in the control of seizures. Barbiturates and benzodiazepines have been shown to enhance GABA receptor function (Choi et al., 1977; Macdonald and Barker, 1978a) but through different allosteric modulatory sites on the GABA receptor (Study and Barker, 1981; Twyman et al., 1989a). Sodium valproate may increase the availability of GABA for release, possibly by inhibiting succinic semialdehyde dehydrogenase, thereby increasing the presynaptic concentration of GABA (Chapman et al., 1982). This mechanism of action for sodium valproate remains unproven. Other agents which may increase GABA availability include GABA prodrugs, such as progabide (Leppik et al., 1987) and GABA transaminase inhibitors, such as γ-vinyl GABA (Metcalf, 1979; Schlechter et al., 1984; Tunnicliff, this volume). The neurosteroids, which include progesterone and progesterone metabolites, can allosterically modulate GABA receptor function (Majewska et al., 1986; Callachan et al., 1987a,b) and have been used in attempts to control seizures in catamenial epilepsy (Mattson et al., 1984). The mechanisms of action of barbiturates, benzodiazepines, and neurosteroids at the GABA$_A$ receptor will be considered in detail. The mechanisms of action of GABA prodrugs, transaminase inhibitors and GABA-mimetics will be considered in the context that they act via a mechanism similar to increased GABA concentration or more potent GABA receptor agonists. GABA prodrugs and transaminase inhibitor agents are discussed further elsewhere in this volume. GABA regulation of its chloride channel will be reviewed first.

GATING OF THE GABA RECEPTOR ION CHANNEL

The GABA receptor is a macromolecular protein composed of a chloride ion selective channel with binding sites at least for GABA, picrotoxin, barbiturates, and benzodiazepines (Olsen, 1982). The GABA receptor appears to be composed of at least three different polypeptide subunits of α, β, and γ chains (Barnard et al., 1987; Prichett et al., 1989). There are at least six α, two β, and two γ isoforms. Although cloned receptors composed of F128a and β subunits open chloride selective channels when exposed to GABA and although these receptors have an increased response in the presence of pentobarbital (Schofield et al., 1987), benzodiazepine sensitivity is not present (Levitan et al., 1988). Full receptor pharmacology required the presence of a γ subunit forming a channel with a likely stoichiometry of 2α2βγ (Prichett et al., 1989).

GABA binds to its receptor to regulate gating (opening and closing) of the chloride ion channel. Two molecules of GABA are necessary for full activation of the native receptor channel. Hydrophobic membrane spanning regions of each of the subunits make up the channel, and the structural and electrical characteristics of the channel pore determine the chloride ion permeability and the conductance of the channel (Barnard et al., 1987).

Single channel recording has revealed that when open, the channel may adopt at least four current conducting levels or conductance states (Bormann et al., 1987). In central murine GABA receptors, a 27–30 pS conductance state (double asterisk, Fig. 1) is the predominant or main-conductance state recorded (Bormann et al., 1987; Macdonald et al., 1989a). The basis for the multiple conductance states remains unknown but the multiple states may reflect the configuration or combination of different receptor subunits or the distribution of charges within the ion channel.

The main-conductance state is responsible for over 90% of the current through the receptor channel, and the single channel gating properties of the main-conductance state of the native GABA receptor in murine spinal cord neurons in culture have been characterized (Macdonald et al., 1989a; Twyman et al., 1990). Single GABA-evoked openings are reduced in the presence of the GABA receptor antagonists bicuculline and picrotoxin (Macdonald et al., 1989a) and these antagonists are also potent convul-

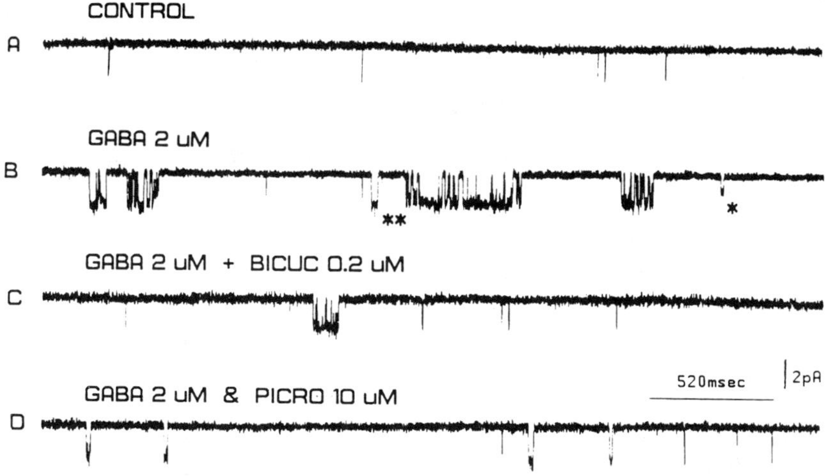

Fig. 1. Single channel GABA receptor currents recorded from patches of mouse spinal cord neurons using an outside-out patch clamp recording configuration. Membranes were voltage clamped at -75 mV and the chloride equilibrium potential was 0 mV. A: Before the exposure of the patch to GABA, rare, brief, spontaneous currents were recorded. Channel openings produced downward deflections of the current recording. B: GABA (2 μM), applied to the patch using pressure ejection micropipettes, produced an increased frequency of channel openings with a predominant current amplitude corresponding to a main-conductance (double asterisk) of about 27–28 picosiemons (pS) and a sub-conductance state (single asterisk) of about 16–19 pS. Openings occurred singly or in groups (bursts) of openings. C: The GABA receptor antagonist, bicuculline (BICUC), reduced the GABA-evoked current. D. Picrotoxin (PICRO) also reduced the GABA-evoked current. (Reproduced from Macdonald et al., 1989a, with permission of the publisher.)

sants (Ajimone-Marsen, 1969). It has been found that binding of GABA increases the probability of channel opening and after the channel opens, it can close and rapidly re-open, to create bursts of openings (Fig. 2). The single channel activity of the main-conductance state has been modelled using a microscopic reaction scheme incorporating two sequential GABA binding sites, three open states and ten closed states (Fig. 3) (Twyman et al., 1990). The channel can open to at least three open states (O_1, O_2, and O_3) that have mean open durations of about 1, 3, or 9 ms, respectively. Bursts, composed of groups of openings (Fig. 2), are generally organized into a series of openings with 1, 3, or 9 ms mean dwell times. Those bursts composed primarily with the 1 ms mean dwell time opening generally have the shortest mean burst duration while those bursts composed primarily with the 9 ms mean dwell time opening have the longest mean burst duration.

Increased concentration of GABA produced increased channel opening and burst frequencies and average open and burst durations of the channel without altering the channel conductance (Fig. 2) (Macdonald et al., 1989a).

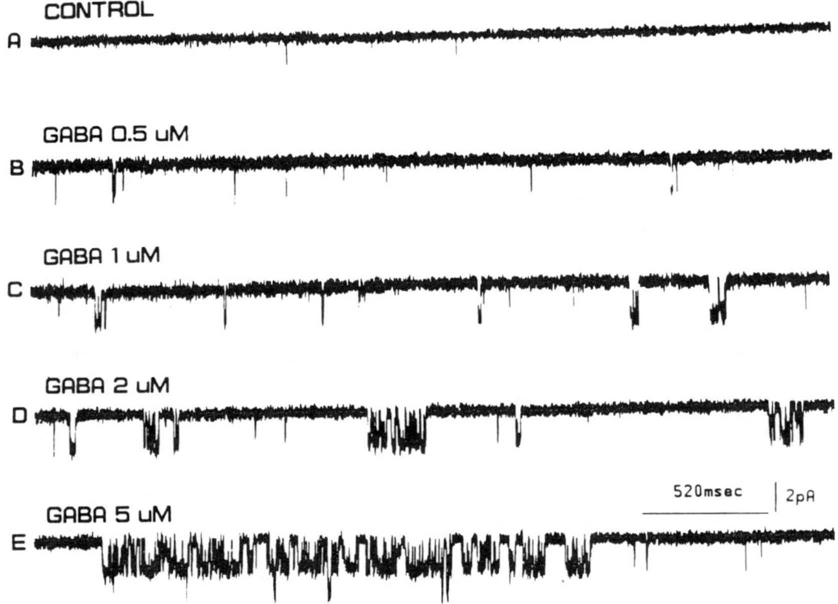

Fig. 2. Single channel GABA-evoked currents were concentration-dependent. Recording conditions were the same as described for Figure 1. **A:** Rare, spontaneous openings were observed before the application of GABA. **B-E:** Application of GABA (0.5–5 μM) produced a concentration-dependent increase in opening and burst frequencies. (Modified from Macdonald et al., 1989a.)

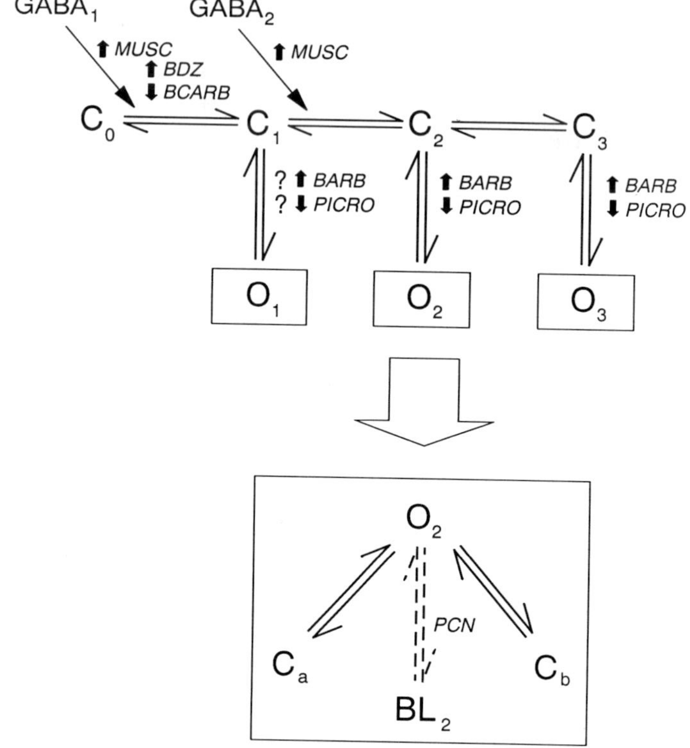

Fig. 3. Microscopic reaction scheme for the GABA receptor main-conductance state shows binding sites for GABA and proposed sites of action of anticonvulsants and convulsants. In this model, GABA binds sequentially to sites GABA$_1$ and GABA$_2$. The channel can exist in multiple open (O) and closed states (C). The open states (O$_1$, O$_2$, and O$_3$) have mean dwell times of 1, 3, and 9 ms, respectively. Each open state can produce a burst of openings by oscillating primarily between itself and two adjacent, distal closed states (C$_a$ and C$_b$) which are shown for O$_2$ in the expanded box. The brief intraburst closures arising from C$_a$ and C$_b$ have mean dwell times of about 0.2 and 2 ms. The observed mean number of openings per burst are 1, 2, and 4 for bursts derived primarily from open states O$_1$, O$_2$, and O$_3$, respectively. The binding of GABA drives the reaction to the right and an increased concentration of GABA would produce an increase. GABA receptor current by an increased frequency of openings nd an increased proportion of openings derived from the longer open states (O$_2$ and O$_3$). Consequently, burst frequency and average burst duration also would be observed with increased concentration. To enhance GABA-evoked current, barbiturates appeared to primarily increase opening transition rates of bound receptors (BARB) thereby prolonging the time spent in open states. The convulsant, picrotoxin (PICRO) acts in a reciprocal fashion compared to the barbiturates. Benzodiazepines modify transition rates (BDZ) or the affinity of the first GABA binding site (GABA$_1$) to increase GABA-evoked channel opening frequency, and thus does not alter average open and burst durations. Convulsant β-carbolines (BCARB) reduce GABA receptor

(Continued)

The increased average open and burst durations have been demonstrated to result not from alteration of dwell times of the open states, but from increased frequency of openings from longer open states and reduced proportion of openings from the shortest open state. From the reaction scheme in the GABA receptor kinetic model (Fig. 3), increased GABA concentration would drive the reaction to the right, thereby producing an increased frequency of opening. With increased GABA concentration, the openings also would be primarily of the longer open states (O_2 and O_3) and thus would produce an increased average channel open duration. Bursts of openings, generally produced by oscillations of open states and their kinetically adjacent but distal closed states (C_a and C_b, boxed inset Fig. 3), also are increased with GABA concentration. An increased frequency and proportion of longer bursts, made up of the longer open states (O_2 and O_3) would effectively increase average burst duration as GABA concentration is increased (Twyman et al., 1990).

Some AEDs can modulate GABA receptor current by regulating single channel properties of the receptor. To enhance GABA receptor current, an agent may 1) increase the channel conductance, 2) increase the channel open and burst frequencies, and/or 3) increase the channel open and burst durations. With a kinetic model of the GABA receptor, the mechanisms of action of AEDs that may act through the GABA receptor may be studied.

MECHANISMS OF AED ACTION AT THE GABA RECEPTOR

Barbiturates

Phenobarbital, in use for epilepsy since 1912, is one of the safest AEDs on the market. Primidone does not have a direct effect on the GABA receptor but it is metabolized to phenobarbital. Pentobarbital may also be used to control seizures, but has a rather low therapeutic index (due to sedation) and is primarily used in the treatment of status epilepticus. Although the barbiturates bind to an allosteric regulatory site on the GABA receptor, the subunit location of the barbiturate binding site is unknown. However, since barbiturates enhance GABA receptor current when the γ subunit has not been added, the binding site must reside on the α and/or the β subunit.

Results from fluctuation analysis have suggested that phenobarbital and pentobarbital prolonged the mean channel open duration of GABA-gated currents without altering the channel conductance (Barker and McBurney,

---

(Figure 3, continued)
currents by a mechanism reciprocally related to anticonvulsant benzodiazepines. GABA prodrugs and GABA transaminase inhibitors increase the availability of GABA to bind at its binding sites. The potent GABA-mimetic muscimol (MUSC) binds to the GABA receptor with increased association rates at both GABA binding sites (GABA$_1$ and GABA$_2$). The convulsant penicillin (PCN) blocks GABA-evoked openings and introduces a new blocked state distal from each of the three open states (shown as BL$_2$ for O$_2$).

1979; Study and Barker, 1981). Direct observation of barbiturate enhanced single channel GABA receptor currents confirmed that these agents did not alter GABA receptor conductance (Fig. 4) (Jackson et al., 1982; Mathers, 1985; Macdonald et al., 1989b; Twyman et al., 1989b). Single channel opening frequency also was unaffected by the barbiturates (Macdonald et al., 1989b). However, analysis of GABA receptor open durations in the presence of clinically revelant free serum therapeutic concentrations of phenobarbital and pentobarbital revealed that they did not alter the dwell times of the open states of the receptor (Macdonald et al., 1989b). These barbiturates shifted the proportion of openings with short dwell times to openings with longer dwell times. Thus, although the duration of the GABA receptor open states were unchanged in the presence of the barbiturates, the average open duration of all openings of the channel was effectively increased. The single channel kinetics also revealed that this effect on increased average open duration was not simply the same effect as increased GABA concentration. The barbiturates appeared to primarily increase transition rates involved in opening of the bound receptor rather than steps involved in the binding of GABA (Fig. 3). Since phenobarbital has not been convincingly shown to enhance GABA binding (Olsen, 1982), these findings suggest that the barbiturates alter the intrinsic gating of the channel.

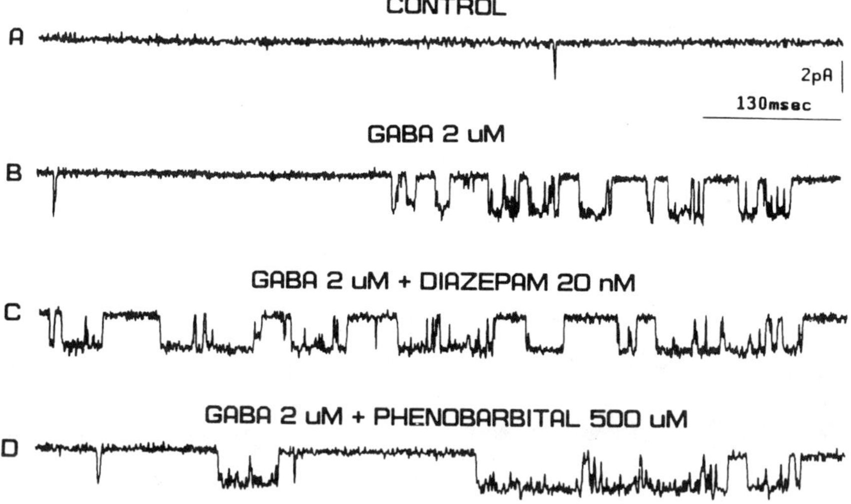

Fig. 4. Single GABA-receptor currents were enhanced by diazepam and phenobarbital. Recording conditions were similar to that described in Figure 1. A: Spontaneous currents in the absence of GABA. B: GABA-evoked bursts of openings. C: GABA-evoked opening and burst frequencies were increased by diazepam. D: Phenobarbital also increased GABA receptor currents by an increased averaged open and burst duration and did not increase the frequency of opening. (Reproduced from Twyman et al., 1989a, with permission of the publisher.)

Further evidence for this was gained from the analysis of burst properties. In the presence of the barbiturates, GABA-evoked bursts were more likely to occur with openings from longer open states. The increased likelihood of reopening of the longer open states effectively prolonged the bursts and thus, prolonged the average burst duration.

At high concentrations, phenobarbital and pentobarbital also can directly activate the GABA receptor (Macdonald and Barker, 1979; Jackson et al., 1982; Akaike et al., 1985), however, the clinical revelance of this direct GABA-mimetic effect is unclear. At anesthetic concentrations, pentobarbital is effective at terminating status epilepticus but also can suppress SRF and reduce calcium current, and thus, calcium mediated transmitter release (Macdonald and McLean, 1986).

Picrotoxin, a convulsant, is known to non-competitively reduce GABA-evoked currents (Macdonald and Barker, 1978b). Both phenobarbital and pentobarbital can displace picrotoxin binding at the GABA receptor although the binding sites of the agents do not completely overlap (Olsen, 1982). Thus, the kinetic mechanism by which picrotoxin reduces GABA-evoked current would be expected to be reciprocal compared to that of the barbiturates. Indeed, single channel recordings have revealed that picrotoxin reduced GABA-evoked average open duration and burst duration (Twyman et al., 1989b). Kinetic analysis of the mechanism by which picrotoxin reduced open and burst durations suggested that picrotoxin reduced transition rates of opening of the bound receptor. This is reciprocally related to the kinetic mechanism by which the barbiturates enhance GABA-evoked currents (Fig. 3). However, in contrast to the barbiturates where GABA receptor opening frequency was unaffected, picrotoxin reduced GABA receptor opening frequency. The basis for this difference is unknown.

Benzodiazepines

Benzodiazepines have a high affinity for the GABA receptor and have been demonstrated to increase the affinity of the receptor for GABA (Olsen, 1982; Skerrit et al., 1982). Results from fluctuation analysis have suggested that diazepam primarily increased GABA receptor opening frequency without altering channel conductance (Study and Barker, 1981). Similar to the barbiturates, single channel recordings confirmed that benzodiazepines do not alter GABA receptor conductance (Fig. 5) (Vicini et al., 1987; Rogers et al., 1988; 1989; Twyman et al., 1989a). If the effect of benzodiazepines to enhance GABA receptor current was purely due to increased affinity of the receptor for GABA, the single channel kinetic properties would be predicted to be similar to an increased concentration of GABA. GABA-evoked channel open and burst frequencies and average channel open and burst durations would be expected to be increased in the presence of a benzodiazepine. Analysis of single channel kinetic properties did not reveal this expected result (Rogers et al., 1988). At clinically revelant, low nanomolar concentrations of diazepam, GABA-evoked open and burst frequencies were increased, but average open and burst durations were unaltered. The increased burst frequency with a change in the mean burst duration in the presence of diazepam is in contradistinction to the marked increase in burst duration and little effect on burst frequency in the presence of phenobarbital (Twyman et al., 1989a). For diazepam, these results could be explained by an increased affinity of GABA at only one of the GABA binding sites (GABA$_1$, Fig. 3) but not both of the GABA binding sites. More specifically, the increased open and burst frequencies with no change in open and burst properties could be explained by an increased association rate or a decreased dissociation rate of only the first binding site. Alteration of these rates for the second binding site would alter significantly the open and burst properties.

The kinetics for the reduction of GABA receptor currents by benzodiazepine receptor antagonists would be expected to be reciprocally related to the mechanism of action of benzodiazepine receptor agonists. Results from this laboratory showed that convulsant β-carbolines such as methyl 6,7-dimethoxy-4-ethyl-β-carboline-3-carboxylate (DMCM) did not alter GABA receptor conductance and average open and burst duration (Rogers et al., 1989). However, open and burst frequencies were reduced. These results suggested that the modulation of GABA receptor single channel kinetics by DMCM could be explained by a reduction of the affinity of GABA binding at the first binding site (GABA$_1$, Fig. 3), but not at both of the GABA binding sites. Thus, the mechanism for the reduction of GABA receptor currents by DMCM was reciprocally related to the mechanism found for the enhancement of GABA receptor currents by diazepam.

In contrast to the barbiturates, the benzodiazepines do not directly activate the GABA receptor at high concentrations. However, at high diazepam concentrations, the GABA-evoked current has been observed to be reduced (Skerrit and Macdonald, 1984). The reduction in current may be due to desensitization or decreased responsiveness of the receptor. This phenomenon is also observed with high concentrations of GABA (Akaike et al., 1987). The basis for this is unknown.

<h2 style="text-align:center">Steroids</h2>

While the intracellular actions of steroid hormones are well known, there recently has been a great deal of interest in a variety of steroids and their derivatives that have been demonstrated to have activity at the GABA receptor. Progesterone metabolites have been known for many years to exert sedative and hypnotic effects and a synthetic steroid, alfaxalone, has been used in the induction of anesthesia (Barker et al., 1987; Callachan et al., 1987a; Cottrell et al., 1987). Recent evidence has demonstrated that certain steroids are potent stereoselective allosteric modulators of the GABA receptor. Some endogenous steroids can interact with GABA receptors at physiological concentrations and thus, may influence central nervous system function during physiological and pathological conditions. It has been speculated that physiologic variability of the levels of pregnane metabolites contribute to the development of stress and anxiety and alter seizure susceptibility (Callachan et al., 1987b). Steroid binding to the GABA receptor has been demonstrated (Gee et al., 1988), but the receptor subunit location remains unknown. Similar to the barbiturates, steroids displace TBPS binding which suggests that steroids and barbiturates have closely associated binding sites. Steroids have been described to potentiate GABA responses in a "barbiturate-like" fashion (Majewska et al., 1986; Barker et al., 1987). Neither steroid nor barbiturate effects are blocked by the benzodiazepine receptor antagonist Ro 15–1788 (Cottrell et al., 1987). Also, both steroids (Barker et al., 1987; Callachan et al., 1987b; Cottrell et al., 1987) and barbiturates at high concentrations can directly activate the GABA receptor. However, some reports have provided evidence that barbiturates and steroids do not have a common site or mechanism of action. The effect of combining steroids and barbiturates on the binding of GABA, TBPS, and benzodiazepines suggests the presence of separate steroid and barbiturate binding sites (Morrow et al., 1990). Direct GABA receptor activation by high concentrations of steroids can be further modulated by low concentrations of barbiturate (Callachan et al., 1987b). In contrast to the barbiturates, structurally different steroids can either potentiate or antagonize GABA responses (Mienville and Vicini, 1989).

Single channel studies of GABA receptor modulation by steroids have shown that the conductance of the receptor is unaltered (Callachan et al., 1987b). Prolongation of mean channel open time has been inferred by fluctuation analysis and marked prolongation of single channel burst duration has been reported (Barker et al., 1987; Callachan et al., 1987b). The kinetic mechanism of steroid modulation of the GABA receptor will have to await detailed analysis of single channel kinetics.

GABA Receptor Agonists

GABA prodrugs, GABA transaminase inhibitors, and GABA-mimetics have been developed with the intention to increase the availability of GABA or a GABA-mimetic agent at the post-synaptic membrane. As discussed above, increased concentration of GABA results in increased chloride current flow via increased channel open and burst frequencies and increased average channel open and burst durations without a change in the intrinsic open state dwell times of the channel. The increased current with increased GABA concentration is not limitless, however, primarily because of concentration induced receptor desensitization and limitations due to characteristics of receptor gating kinetics.

GABA receptor agonists increase GABA receptor current presumably by acting through one or both of the bicuculline-sensitive GABA binding sites. Muscimol, a plant alkaloid and a potent GABA receptor agonist (Mathers and Barker, 1981), has been used in the treatment of animal models of epilepsy (Löscher, 1982), but muscimol can also precipitate seizures as a toxic side effect (Pedley et al., 1979). Muscimol evokes GABA receptor chloride currents with single channel conductances similar to those evoked by GABA (Jackson et al., 1982; Twyman et al., manuscript submitted).

Results from this laboratory indicate that muscimol evoked bursting currents similar to GABA, but at the same concentrations, open and burst frequencies and average open and burst durations were greater when evoked by muscimol as compared to GABA. Open state dwell times, however, were similar to those of GABA. Single channel kinetic modelling revealed that the basis for this could be explained by muscimol binding not at one of the GABA binding sites, but at each of the GABA receptor binding sites with greater association rates. From the kinetic reaction scheme, increased association rates at both of the GABA binding sites ($GABA_1$ and $GABA_2$, Fig. 3) would increase the frequency and duration of longer openings and bursts. The increased association rates provided the basis for the observed increased affinity of muscimol for the GABA receptor as compared to GABA. Therefore, the potent GABA-mimetic action of muscimol could be explained simply by its greater association rates for both of the GABA receptor binding sites. It remains to be seen whether or not other GABA-mimetic agents act via a similar mechanism, but it seems plausible that they do.

GABA Receptor Chloride Channel Conductance

All agents discussed thus far have been shown to regulate the GABA receptor gating characteristics. No agent has been demonstrated to regulate the chloride channel conductance as a mechanism for modulating chloride current flow through the receptor. Open channel blockers have been demonstrated to gain entry into ion channels and physically block current flow. Current flow is usually completely occluded when the channel is "blocked" but when the channel is unblocked, the channel conductance is unaltered. For the GABA receptor, evidence for open channel blockade by penicillin has been found (Chow and Mathers, 1986; Twyman et al., 1991).

Penicillin has been shown to reduce synaptic inhibition and in toxic doses, has produced seizures *in vivo* (Raichle et al., 1971). GABA-evoked cellular currents in the presence of penicillin are reduced in amplitude and are prolonged in duration (Macdonald and Barker, 1978b). Single channel recordings from this laboratory have revealed that the basis for the reduced current amplitude is primarily due to a reduction of average channel duration and not by reduction of single channel conductance. The basis for the observed prolongation of GABA-evoked cellular currents in the presence of penicillin was primarily due to prolongation of burst durations. The single channel kinetics of the reduction of open state duration and prolongation of burst duration was consistent with open channel blockade of the GABA receptor. In the GABA receptor microscopic reaction scheme, penicillin introduced a distal blocked closed state (BL, Fig. 3) for each of the three open states. Penicillin, a negatively charged molecule at physiological pH, must therefore interact with positively charged proteins within the channel to intermittently occlude chloride ion flow through the channel.

It appears that it would be unlikely that agents would be developed that could significantly modify GABA receptor conductance. Such an agent would have to alter charge distribution within the channel pore or somehow change the receptor subunit structure or configuration. As can be seen from the above review, it is far easier to regulate gating characteristics through allosteric regulatory sites than to modify the structural integrity of the channel.

CONCLUSIONS

The GABA receptor is highly regulated and several AEDs take advantage of preexisting allosteric regulatory sites to enhance GABA receptor function. Understanding the mechanisms of action of these AEDs can help further their utility and the design of newer AEDs. This is true, particularly in light of recent findings of multiple types of receptor subunit isoforms and the likelihood of regionally specific combinations of subunits. With the understanding of how agents regulate GABA receptors and their subtypes, the potential for the design of drugs to act at site specific GABA receptors is great.

REFERENCES

Ajimone-Marsen C: Acute effects of topical epileptogenic agents. In: Jasper HN, Ward AA and Pope A, eds., "Basic Mechanisms of the Epilepsies." Boston: Little, Brown & Co. 1969, pp. 299–319.

Akaike N, Hattori K, Inomata N, Oomura Y: γ-Aminobutyric-acid- and pentobarbitone-gated chloride currents in internally perfused frog sensory neurones. J Physiol (Lond) 360: 367–386, 1985.

Akaike N, Inomata N, Tokutomi N: Contribution of chloride shifts to the fade of γ-aminobutyric acid-gated currents in frog dorsal root ganglion cells. J Physiol (Lond) 391: 219–231, 1987.

Barker JL, Harrison NL, Lange GD, Owen DG: Potentiation of γ-aminobutyric-acid-activated chloride conductance by a steroid anesthetic in cultured rat spinal neurones. J Physiol (Lond) 386: 485–501, 1987.

Barker JL, McBurney RM: GABA and glycine may share the same conductance channel on cultures mammalian neurons. Nature 277: 234–236, 1979.

Barnard EA, Darlison MG, Seeburg P: Molecular biology of the GABA$_A$ receptor: the receptor/channel superfamily. Trends Neurosci 10: 502–509, 1987.

Bormann J, Hamill OP, Sakmann B: Mechanism of anion permeation through channels gated by glycine and γ-aminobutyric acid in mouse cultured spinal neurons. J Physiol (Lond) 385: 243–286, 1987.

Callachan H, Lambert JJ, Peters JA: Modulation of the GABA$_A$ receptor by barbiturates and steroids. Neurosci Lett Suppt. 29: S21, 1987a.

Callachan H, Cottrell GA, Hather NY, Lambert JJ, Nooney JM, Peters JA: Modulation of the GABA$_a$ receptor by progesterone metabolites. Proc R Soc (Lond) [Biol] B231: 359–389, 1987b.

Chapman AG, Riley K, Evans MC, Meldrum, BS: Acute effects of sodium valproate and γ-vinyl GABA on regional amino acid metabolism in the rat brain. Neurochem Res 7: 1089–1105, 1982.

Choi DW, Farb DH, Fischbach GD: Chlordiazepoxide selectively augments GABA action in spinal cord cell cultures. Nature 269: 342–344, 1977.

Chow P, Mathers D: Convulsant doses of penicillin shorten the lifetime of GABA-induced channels in cultured central neurones. Br J Pharmacol 88: 541–547, 1986.

Cottrell GA, Lambert JJ, Perters JA: Modulation of $GABA_A$ receptor activity by alphaxalone. Br J Pharmacol 98: 491–500, 1987.

Coulter DA, Huguenard JR, Prince DA: Characterization of ethosuximide reduction of low-threshold calcium current in thalamic neurons. Ann Neurol 25: 582–589, 1989a.

Coulter DA, Huguenard JR, Prince DA: Calcium currents in rat thalamocortical relay neurones; kinetic properties of the transient, low-threshold current. J Physiol (Lond) 414: 587–604, 1989b.

Courtney KR, Etter EF: Modulated anticonvulsant block of sodium channels in nerve and muscle. Eur J Pharmacol 88: 1–9, 1983.

Croucher MJ, Collins JF, Meldrum BS: Anticonvulsant action of excitatory amino acid antagonists. Science 216: 899–901, 1982.

Gee KW, Bolger MB, Brinton RE, Coirini H, McEwen BS: Steroid modulation of the chloride iontophore in rat brain: structure-activity requirements, regional dependence and mechanism of action. J Pharmacol Exp Ther 246: 803–812, 1988.

Gross RA, Macdonald RL: Barbiturates and nifedipine have different and selective effects on calcium currents of mouse DRG neurons in culture: A possible basis for differing clinical actions. Neurology 18: 443–451, 1987.

Gross RA, Macdonald RL: Differential actions of pentobarbitone on calcium current components of mouse sensory neurones in culture. J Physiol (Lond) 405: 187–203, 1988.

Gross RA, Kelly KM, Macdonald RL: Ethosuximide and dimethidione selectively reduce calcium currents in cultured sensory neurons by different mechanisms. Neurology 39 (Supp/1):412, 1989.

Jackson MB, Lecar H, Mathers DA, Barker JL: Single channel currents activated by γ-aminobutyric acid, muscimol, and (-)pentobarbital in cultured mouse spinal neurons. J Neurosci 2: 889–894, 1982.

Kelly KM, Gross RA and Macdonald RL: Valproic acid selectively reduces the low-threshold (T) calcium current in rat nodose neurons. Neurosci Lett 116: 233–238, 1990.

Leppik IE, Dreifuss FE, Porter R, Bowman T, Santilli N, Jacobs M, Crosby C, Cloyd J, Stackman J, Graves N, Sutula T, Welty T, Vickery J, Brundage R, Gates J, Gumnit RJ, Gutierrez: A controlled study of progabide in partial seizures: methodology and results. Neurology 37: 963–968, 1987.

Levitan ES, Blair LAC, Dionne VE, Barnard EA: Biophysical and pharmacological properties of cloned $GABA_A$ receptor subunits expressed in *Xenopus oocytes*. Neuron 1: 773–781, 1988.

Löscher W: Comparative assay of anticonvulsant and toxic potencies of sixteen GABAmimetic drugs. Neuropharmacology 21: 803–810, 1982.

Macdonald RL: Antiepileptic drug actions. Epilepsia 30 (Supp/1):S19–S28, 1989.

Macdonald RL, Meldrum BS: General principles—principles of antiepileptic drug action. In Levy R, Mattson R, Meldrum B, Penry JK, Dreifuss FE (eds): "Antiepileptic Drugs," 3rd ed. New York: Raven Press, 1989, pp. 59–83.

Macdonald RL, Barker JL: Benzodiazepines specifically modulate GABA-mediated postsynaptic inhibition in cultured mammalian neurones. Nature 271: 563–564, 1978a.

Macdonald RL, Barker JL: Specific antagonism of GABA-mediated postsynaptic inhibition in cultured mammalian spinal cord neurons: a comon mode of convulsant action. Neurology 28:325–330, 1978b.

Macdonald RL, Barker JL: Anticonvulsant and anesthetic barbiturates: different postsynaptic actions in cultured mammalian neurons. Neurology 29: 432–447, 1979.

Macdonald RL, McLean MJ: Anticonvulsant Drugs: Mechanisms of Action. In Delgado-Escueta AV, Ward AA Jr, Woodbury DM, Porter RJ (eds): "Antiepileptic drugs: Mechanisms of action." (Advances in Neurology, Vol. 44.) New York: Raven Press, 1986, pp. 713–736.

Macdonald RL, Rogers CJ, Twyman RE: Kinetic properties of the GABA$_A$ receptor main-conductance state of mouse spinal cord neurons in culture. J Physiol (Lond) 410: 479–499, 1989a.

Macdonald RL, Rogers CJ, Twyman RE: Barbiturate modulation of kinetic properties of GABA$_A$ receptor channels in mouse spinal neurons in culture. J Physiol (Lond) 417: 483–500, 1989b.

Majewska MD, Harrison NL, Schwartz RD, Barker JL, Paul SM: Steroid hormone metabolites are barbiturate-like modulators of the GABA receptor. Science 232: 1004–1007, 1986.

Mathers DA: Spontaneous and GABA-induced single channel currents in cultured murine spinal cord neurons. Can J Physiol Pharmacol 63: 1228–1233, 1985.

Mathers DA, Barker JL: GABA and muscimol open ion channels of different lifetimes on cultured mouse spinal cord cells. Brain Res 204: 242–247, 1981.

Mattson RH, Cramer JA, Caldwell BV, Siconolfi BC: Treatment of seizures with medroxyprogesterone acetate: preliminary report. Neurology 34: 1255–1258, 1984.

Mayer ML, Westbrook GL: The physiology of excitatory amino acids in the vertebrate central nervous system. Prog Neurobiol 28: 197–276, 1987.

McLean MJ, Macdonald RL: Sodium valproate, but not ethosuximide, produces use- and voltage-dependent limitation of high frequency repetitive firing of action potentials of mouse central neurons in cell culture. J Pharmacol Exp Ther 237: 1001–1011, 1986.

Meldrum BS, Croucher MJ, Czuczwar SJ, Collins JF, Curry K, Joseph M, Stone TW: A comparison of the anticonvulsant potency of (+/-) phosphonopentanoic acid and (+/-) 2-amino-7-phosphonoheptanoic acid. Neuroscience 9: 925–930, 1983a.

Metcalf BW: Inhibitors of GABA metabolism. Biochem Pharmacol 28: 5–17, 1979.

Meyer FB, Anderson RE, Sundt, Jr. TM, Sharbrough FW: Selective central nervous system calcium channel blockers—a new class of anticonvulsant agents. Mayo Clin Proc 61: 239–247, 1986.

Mienville JM, Vicini S: Pregnenolone sulfate antagonizes GABA$_A$ receptor-mediated currents via a reduction of channel opening frequency. Brain Res 489: 190–194, 1989.

Morrow AL, Pace JR, Prudy RH, Paul SM: Characterizations of steroid interactions with the GABA receptor-gated ion channel: evidence for multiple steroid recognition sites. Mol Pharmacol 37: 263–270, 1990.

Nowycky MC, Fox AP, Tsien RW: Three types of neural calcium channels with different calcium agonist sensitivity. Nature 316: 440–443, 1985.

Olsen RW: Drug interaction at the GABA receptor-iontophore complex. Annu Rev Pharmacol Toxicol 22: 245–277, 1982.

Overweg J, Binnie CD, Meijer JWA, Meinardi H, Nuijten STM, Schmaltz S, Wauquier A: Double-blind placebo-controlled trial of flunarizine as add-on therapy in epilepsy. Epilepsia 25: 217–222, 1984.

Pedley TA, Horton RW, Meldrum BS: Electroencephalographic and behavioural effects of a GABA agonist (muscimol) on photosensitive epilepsy in the baboon, *Papio papio*. Epilepsia 20: 409–416, 1979.

Prichett DB, Sontheimer H, Shivers BD, Ymer S, Kettenmann H, Schofield PR, Seeburg PH: Importance of a novel GABA$_A$ receptor subunit for benzodiazepine pharmacology. Nature 338: 582–584, 1989.

Raichle ME, Kult H, Louis S, McDowell F: Neurotoxicity of intravenously administered penicillin G. Arch Neurol 25: 232–239, 1971.

Rogers CJ, Twyman RE, Macdonald RL: Diazepam does not alter the gating kinetics of GABA receptor channels. Soc Neurosci Abst 14: 642, 1988.

Rogers CJ, Twyman RE, Macdonald RL: The benzodiazepine diazepam and the beta-carboline DMCM modulate GABA$_A$ receptor currents by opposite mechanisms. Soc Neurosci Abst 15: 1150, 1989.

Schlechter PJ, Hanke NFJ, Grove J, Huebert N, Sjoerdsma A: Biochemical and clinical effects of β-vinyl GABA in patients with epilepsy. Neurology 34: 182–186, 1984.

Schofield PR, Darlison MG, Fujita N, Burt DR, Stephenson FA, Rodriguez H, Rhee LM, Ramachandran J, Reale V, Glencorse TA, Seeburg PA, Barnard EA: Sequence and functional expression of the GABA$_A$ receptor shows a ligand-gated receptor super-family. Nature 328: 221–227, 1987.

Skerrit JH, Macdonald RL: Benzodiazepine receptor ligand actions on GABA-responses. β-carbolines, purines. Eur J Pharmacol 101: 135–141, 1984.

Skerrit JH, Willow M, Johnston GAR: Diazepam enhancement of low affinity GABA binding to rat brain membranes. Neurosci Lett 29: 63–66, 1982.

Study RE, Barker JL: Diazepam and (+/-) pentobarbital: Fluctuation analysis reveals different mechanisms for potentiation of γ-aminobutyric acid responses in cultured central neurons. Proc Natl Acad Sci USA 78: 7180–7184, 1981.

Twyman RE, Green RM, Macdonald RL: Kinetics of open channel block of GABA$_A$ receptor channels by penicillin. J Physiol (Lond). In press, 1991.

Twyman RE, Rogers CJ, Macdonald RL: Differential mechanisms for enhancement of GABA by diazepam and phenobarbital: A single channel study. Ann Neurol 25: 213–220, 1989a.

Twyman RE, Rogers CJ, Macdonald RL: Pentobarbital and picrotoxin have reciprocal actions on single GABA-Cl channels. Neurosci Lett 96: 89–95, 1989b.

Twyman RE, Rogers CJ, Macdonald RL: Intraburst kinetic properties of the GABA$_A$ receptor main conductance state of mouse spinal cord neurons in culture. J Physiol (Lond) 423: 193–220, 1990.

Vicini S, Mienville JM, Costa E: Actions of benzodiazepine and beta-carboline derivatives on GABA-activated Cl channels recorded from membrane patches of neonatal rat cortical neurons in culture. J Pharmacol Exp Ther 243: 1195–1201, 1987.

Willow M, Catterall WA: Inhibition of binding of [3H]batrachotoxinin A 20-α-benzoate to sodium channels by the anticonvulsant drugs diphenylhydantoin and carbamazepine. Mol Pharmacol 22: 627–635, 1982.

Willow M, Kuenzel EA, Catterall WA: Inhibition of voltage-sensitive sodium channels in neuroblastoma cells and synaptosomes by the anticonvulsant drugs diphenylhydantoin and carbamazepine. Mol Pharmacol 25: 228–234, 1984.

5
GABA Neurotransmitter Activity in Human Epileptogenic Brain

GODFREY TUNNICLIFF AND BEAT U. RAESS

Laboratory of Neurochemistry (G.T.) and Department of Pharmacology and Toxicology (B.U.R.), Indiana University School of Medicine, Evansville, IN 47712

INTRODUCTION

It is well accepted that γ-aminobutyric acid (GABA) plays a crucial role in maintaining inhibitory influences in the central nervous system. The primary vehicle of this inhibition is the $GABA_A$-benzodiazepine receptor complex which mediates the transmitter-evoked Cl^- flux into specific neurons. Details of the mechanisms involved are reviewed by Krnjević (Chapter 3, this volume).

There is little doubt that when GABA function is compromised under experimental conditions, a large number of neurons undergo an abnormal synchronized discharge. This leads to the characteristic generalized convulsions which resemble grand mal seizures in humans. Such observations naturally led to the concept that the integrity of the inhibitory capacity of the brain of epileptics is diminished; and since GABA is the principal inhibitory neurotransmitter, it is an attractive idea that the GABA system is dysfunctional in the epileptogenic brain.

One of the first methods used to alter GABA metabolism in experimental animals was the inhibition of glutamate decarboxylase (GAD) by a series of convulsant hydrazides that interfered with pyridoxal 5'-phosphate involvement in the catalytic formation of GABA (Killam and Bain, 1957). The onset of convulsions coincided with a sharp decrease in brain GABA content and the inhibition of GAD. Since then, many studies using a variety of convulsants in several different species have demonstrated that inhibition of GAD invariably leads to seizures (Tunnicliff, 1990).

A second way to reduce GABA function is to administer antagonistic drugs which bind at the $GABA_A$ receptor-chloride channel complex and thus interfere directly or indirectly with the Cl^- flux into the cell which, of course, is the basis of the inhibitory action of GABA. Examples of drugs which fall into this category are picrotoxin and bicuculline.

If, indeed, the GABA system is not functioning normally in the human epileptogenic brain, what are the possible reasons? It could be, for instance, because GABA activity is reduced. If this is the case, a likely explanation is that the lower enzyme activity is simply a reflection of a reduced number of GABAergic interneurons. An alternative explanation for a possible deficit in GABA-mediated neurotransmission in the brain of epileptic patients, is a deficit in the number of GABA_A receptors. Again, this could be a consequence of a loss of postsynaptic entities normally under the control of GABA-releasing nerve cells. This review summarizes and discusses results, as well as potential new strategies, from a number of investigations which have been carried out in an attempt to answer these questions.

METHODS USED TO STUDY GABA IN HUMANS
GABA Content in CSF

Several studies have used cerebrospinal fluid (CSF) in an attempt to establish that alterations in GABA levels occur in the central nervous system of epileptic patients. The chief advantage of using CSF is, of course, that it is a relatively non-invasive method of obtaining samples for analysis. Because the CSF bathes the brain and spinal cord, it may be able to provide data on the biochemical state of the central nervous system. In fact, it has been established that GABA levels in the CSF correlate reasonably well with concentrations in the central nervous system (Böhlen et al., 1979; Grove et al., 1982). As with any human studies, differences in age, sex, drug status, and time of sample collection should, ideally, be minimized (Hare et al., 1980). Further, care should be taken in storing the samples since it has been reported that GABA levels can increase in CSF that has not been cooled adequately. This increase is probably the result of homocarnosine breakdown (Grossman et al., 1980).

Wood et al. (1979) collected CSF by lumbar puncture from 21 epileptic patients with intractable seizures and who were receiving various anticonvulsants. GABA levels were measured by a relatively new fluorescence procedure after separation on an ion-exchange column and subsequent derivitization by o-phthalaldehyde. The amino acid levels were compared with those from 20 healthy volunteers. Whereas the volunteers had a mean GABA concentration of 239 ± 21 pmol per ml, the patients with epilepsy had a mean value of 139 ± 12 pmol per ml of CSF. Thus there was a clear-cut reduction (58%) in GABA levels in the CSF of the epileptic patients. In a previous study in which GABA levels were measured by a radioreceptor assay in the CSF of twelve patients with intractable epilepsy, the mean GABA content of the control group was 225 ± 39 pmol per ml (Enna et al., 1977). The mean GABA concentration in the CSF of the epileptic patients was 159 ± 17 pmol per ml, representing a 41% decrease. However, the scatter in the data, particularly of the control group meant that the reduction was not statistically significant. For this investigation, unfortunately, instead of healthy volunteers acting as the control group, eleven patients were chosen who had not been diagnosed as having epilepsy, but instead had other neurological disorders. This may well have contributed to the high variability in GABA content since it has been reported that reduced GABA levels occur in the CSF of other neurological patients (Enna et al., 1980).

Several other studies have been published in which the content of GABA in the CSF of epileptic patients has been determined. Most, however, do not support the idea that GABA concentrations are abnormally low in patients with seizure disorders. For example, Kuroda et al. (1982) did not find evidence of reduced GABA concentration in the CSF of epileptic patients. In contrast, however, in the CSF of a group of young patients who had experienced febrile convulsions, Löscher et al. (1981) noted a significant decrease in GABA levels. The mean GABA concentration in the CSF of the children who had undergone seizures no longer than two hours previously was 134 pmol/ml, whereas the mean value from a control group of patients was 210 pmol/ml. These results, on the other hand, were not confirmed later in a similar study (Knight et al., 1985).

In a more recent study, Crawford and Chadwick (1987) assayed the amino acid content in the CSF from 69 epileptic patients and 51 controls, but noted no differences in GABA content between the two groups. In addition, two other investigations yielded similar findings (Araki et al., 1988; Pitkanen et al., 1989).

GABA Content, Metabolism, and Receptors in Brain

GABA content. Because data from the analysis of the CSF is open to interpretation, it is obviously more satisfactory to measure GABA metabolism in the brain itself and, in fact, a number of studies using central nervous tissue have been published. Using brain tissue removed from 16 patients who had undergone neurosurgical treatment for intractable seizures, Van Gelder et al. (1972) measured the content of several amino acids. Both GABA and aspartate were considerably lower in the cerebral cortex (temporal and frontal lobes) at sites peripheral to the focus as well as at the focus. At the same time, both glutamate and taurine were reduced at the focus, whereas glycine content appeared to be markedly elevated at this site. At first glance, these results seem to implicate GABA in the epilepsy of these patients. Unfortunately, no control tissue was analyzed concurrently. Instead, these investigators compared their data with values obtained from the literature. This has to throw doubt on the validity of the observations. In an animal model of epilepsy, seizures were induced by the application of cobalt powder to the motor cortex of the cat (Van Gelder and Courtois, 1972). The adjacent epileptogenic area was found to be low in GABA, glutamate and aspartate, whereas glutamine and glycine levels were above normal. As with the previous study with human epileptic tissue, these experiments with cats suffer from doubts about the adequacy of controls. The claims that GABA concentrations were lower in the epileptic animals was based on GABA levels in just one animal which did not undergo seizures. Admittedly though, reduction in GABA content did appear to correlate with the severity of seizures. Also worth noting is that a few years later other investigators reported the amino acid content of brain tissue removed from epileptic subjects (Perry et al., 1975). In that report it is clear that at the foci of maximum seizure activity, no reduction in GABA concentration was found. This time adequate control nervous tissue from non-epileptic patients was used. In addition, the apparent changes in taurine and glycine seen by Van Gelder et al. (1972) were not confirmed.

Another investigation reported GABA content in brain tumors from twenty-one non-epileptic patients was compared with tumors from twenty epileptic patients (Schmidt et al., 1984). Although GABA concentrations tended to be higher in the epileptic tissue, no statistically significant differences emerged. Obviously results from this type of study are difficult to evaluate because several different types of neoplasms were involved. The results of Sherwin et al. (1988) appear to confirm the previous two reports. These investigators measured amino acid levels in the cerebral cortex of epileptic patients after seizure surgery. The tissue obtained was from cortex which displayed epileptic activity. Control tissue was from patients who showed epileptic activity only in the hippocampus or amygdala. No differences in GABA concentration were found between the two groups. The excitatory amino acids, glutamate and aspartate, however, were elevated in the tissue from patients exhibiting cortical epileptogenic activity.

In a prior report, however, evidence was presented that in certain patients with epilepsy, GABA content in epileptogenic tissue was diminished (Lloyd et al., 1984). GABA were determined in cortical material removed from thirty one patients being surgically treated for intractable seizures. Control values were obtained from non-epileptic areas in the same patients. In six of the nineteen patients without tumors, a conclusive decrease in GABA concentration was observed, with the remainder showing no alterations or, indeed, showing an increase (two patients). In twelve people with tumors present, eight were identified as having a reduction in cortical GABA levels, and no changes were found in four patients.

Loss of GABA interneurons appears to occur in a rare hereditary neurological disorder called neuronal ceroid-lipofuscinosis. One remarkable feature of the dominant form of this disorder is the presence of grand mal seizures (Boehme et al., 1971). These patients undergo an almost complete degeneration of cortical stellate cells, probably as a result of the accumulation of a lipopigment (Braak and Goebel, 1978). The stellate neurons are known to be GABA-containing, thus this presumed loss of cortical GABAergic inhibition could well be related to the development of generalized seizures in these patients. In fact, a reduction in brain GABA levels in patients with this neurological disorder has been reported (Zeman, 1971).

Glutamate decarboxylase activity. Another measure of GABA metabolism is the activity of GAD, the enzyme responsible for GABA biosynthesis. McGeer et al. (1971) obtained biopsy tissue from three patients diagnosed with epilepsy and measured enzyme activity in several parts of the telencephalon. The results were compared with GAD activity from the same regions of postmortem brain from non-epileptics. Of the three patients studied, two were male and there was no evidence that the enzyme activity from their brain samples was different from control values. However, the third patient was a thirteen-year-old girl and in the four regions of her cerebral cortex examined, there was a consistent marked reduction (ca. 50%) in GAD activity. The significance of these results is extremely difficult to assess owing to the small sample number.

From a group of nineteen epileptic patients undergoing surgical treatment, GAD activity was assayed in cortical tissue identified as epileptogenic or affected by the lesion according to EEG analysis (Lloyd et al., 1981). Enzyme activity was compared to control tissue that had been removed from unaffected areas of the cortex. Nine patients exhibited obviously reduced enzyme activity, although the authors stated that values obtained from one of these patients was not statistically significant. The enzyme activity from nine other patients did not differ from control values, and one patient exhibited elevated GAD activity. In five other patients where tumors had been identified, four showed reduced enzyme activity.

In a study referred to previously, some epileptic patients had reduced GABA levels in the cerebral cortex (Lloyd et al., 1984). In the same study, glutamate decarboxylase activity was also determined. Forty-eight percent of twenty three patients had a definite reduction in cortical enzyme activity. In eleven patients who had tumors, nine were found to have decreased enzyme activity (82%). In a follow-up study, a much larger group of epileptic patients (45) were involved (Lloyd et al., 1986). Enzyme was measured in epileptogenic cortical tissue and was compared with normal tissue from the same patients, four of whom had brain tumors. GAD activity in the pathological tissue was noticeably lower ($P < 0.01$) than in the control tissue (Lloyd et al., 1986).

Rougier et al. (1984) examined cerebral cortical GAD activity at the focus of maximum seizure activity in six patients with intractable epilepsy. The enzyme activity was much reduced in five of the subjects, whereas it was elevated in one patient. In four other epileptic patients, GAD activity was substantially decreased in the hippocampus. If a reduction in GAD activity does occur in certain types of human epilepsy, this would merely confirm results from animal studies. For example, monkeys made epileptic by the application of alumina gel to the cerebral cortex underwent a noticeable reduction in GAD-positive terminals at the epileptogenic sites (Ribak et al., 1979; Houser et al., 1986). However, not all studies using human brain tissue have concluded that reductions in GAD activity accompany epilepsy. For instance, in surgical specimens of brain tumors from various patients, GAD activity was not found to differ between epileptic and non-epileptic patients (Schmidt et al., 1984). Similarly, Sherwin et al. (1984) reported that when GAD activity was measured in epileptic foci of the neocortex, no changes were observed compared to the enzyme from peripheral cortical tissue not showing epileptic activity.

Other studies not only do not find a reduction in GAD activity, they conclude an *elevation* of enzyme occurs in some patients. In resected hippocampal tissue from patients whose seizures originated in this part of the telencephalon, no reduction in GAD-containing nerve cells or nerve terminals could be found (Babb et al., 1989). There was, however, a reduction in the total number of cells, strongly suggesting that in fact there was a relative increase in GABAergic cells. Moreover, in six of the patients studied, the anterior hippocampus initiated the seizures. In these subjects, the number of anterior hippocampal GAD-containing neurons was greater than the same types of neurons in the posterior hippocampus. These investigators concluded that a "hyperinnervation" by GABA releasing cells occurred in their patients and proposed that each GABA neuron would make contact with, and thus inhibit, a greater number of neurons than usual. Consequently, when these cells did eventually fire synchronously, this would be more likely to lead to uncontrolled, paroxysmal discharges.

In support of this apparent increase in GABA activity in human epilepsy, are the experiments of Feldblum et al. (1990) in which quantitative in situ hybridization to GAD mRNA was used in rats with induced temporal lobe epilepsy. One week after lesions were made in the hippocampus by kainic acid treatment, there was a marked increase in GAD mRNA in the dentate, the CA1 of the hippocampus, the parietal cortex and the thalamus. Furthermore, in rats made seizure-susceptible after kainic acid treatment, Marksteiner and Sperk (1988) reported a long-term increase in GAD.

Receptors. Another measure of GABA function is the amount of binding to the GABA$_A$-benzodiazepine receptor complex. In a similar monkey model of epilepsy to that described above (Ribak et al., 1979), a marked reduction in GABA receptor binding in the cortical foci area has been reported (Bakay and Harris, 1981). In humans, Lloyd et al. (1984) examined [^{3}H]GABA binding to cortical homogenates from epileptic and non-epileptic patients. Of those patients without tumors, 18 of 26 displayed reduced binding, and of those seven patients with tumors, three showed a decrease. Thus of the epileptic patients studied, 64% exhibited a reduction in [^{3}H]GABA binding. On the other hand, when GABA binding in tumor tissue excised from epileptic patients was compared to tumor tissue from non-epileptics, a substantial increase was observed (Schmidt et al., 1984). Taking a different approach by studying the benzodiazepine binding site, Sherwin et al. (1986) did not show changes in [^{3}H]flunitrazepam receptor binding in homogenates of human epileptic foci. Thirteen patients with partial seizures were employed. Cortical tissue from epileptic foci was compared with tissue from non-spiking cortex, but neither B$_{max}$ nor K$_d$ differed between the sites.

In Vivo Localization of GABA Function

Non-invasive methods are now becoming available to monitor various aspects of GABA neurotransmission in the intact brain. For example, positron emission tomography (PET) studies have been used to monitor benzodiazepine receptor numbers in animals undergoing experimental seizures (Paul et al., 1978). More recently, in human epileptic foci a reduction in benzodiazepine receptor binding has been reported (Savic et al., 1988). Ten patients with partial epilepsy were injected with[^{11}C]Ro15 1788 (flumazil), a benzodiazepine receptor antagonist, and 15 minutes later the imaging was performed. Both total number of receptors and the affinity of the ligand for the receptors were measured by the "equilibrium approach" of Farde et al. (1986). A group of healthy subjects served as controls. In all three areas of the epileptic cerebral cortex studied, an impressive decrease in the number of benzodiazepine receptors was observed, though no obvious differences in receptor-ligand affinities were found.

In a single-photon emission computer tomography (SPECT) imaging procedure using ten patients with focal, partial epilepsy, Bartenstein et al. (1989) measured benzodiazepine receptor binding after administration of the ^{123}I-labeled Ro 16–0154, a flumazil analogue and inverse agonist. Two hours after the administration of labeled ligand, imaging demonstrated that a reduced amount of receptor binding occurred in seven of the patients, in that part of the cortex deemed to be at the epileptic focus. The reduction in binding was in relation to the binding observed on the contralateral side to the focus. One patient showed no reduced binding, whereas two patients appeared to exhibit an increase in binding of the ligand. Interestingly, this study also found an excellent correlation between cerebral blood flow and the decrease in benzodiazepine receptor binding.

Summary

If GABA is to be implicated in any type of epileptogenic activity, it is germane that the nature and localization of the dysfunction be known. This may include understanding receptor characteristics and densities, and neurotransmitter synthesis and catabolism at the focus or areas projecting to the focus. Some earlier studies made use of CSF by measuring GABA levels. This measurement may have some application since it is relatively easy to obtain samples and it may also in some way be a reflection of the overall metabolic state before, during or after an epileptic episode. Unfortunately, this approach is unlikely to result in any meaningful insights as to GABA function in the epileptic foci. The inherent disadvantage with this concept is probably the prime reason for the inconsistencies found in the literature and reviewed in this chapter.

A more productive strategy to pursue some of these questions undoubtedly has to involve the direct measurement of the affected brain areas, whether this be in terms of receptor function or, for instance, GAD activity. Several such studies have been done using either postmortem or surgically removed brain tissue. It is recognized that this approach has several drawbacks and may account for many of the inconsistent data reported. Newer, more promising techniques are becoming available which offer the prospect for monitoring neurotransmitter function, as well as related glucose metabolism and bloodflow measurements in situ by non-invasive methods such as PET and SPECT.

Overall, most studies involving GABA levels, receptor densities and GAD activities show considerable disagreement. As has been pointed out by Babb et al. (1989), it may well be that many of the discrepancies can be explained by the lack of appropriate control tissues. For example, there may be changes in cell numbers due to cell loss or functional cell death at the foci, a situation which has typically not been addressed. Indeed, several studies provide good evidence that neuronal loss accompanies epileptic activity (Mouritzen-Dam, 1980; Babb et al., 1984; 1989). A general consensus is emerging that this cell loss is caused by excessive Ca^{2+} entry and overload, with detrimental consequences (Pumain et al., 1985; Gloor, 1989). It is hoped that elimination of these uncertainties by the use of more powerful techniques will lead to more effective, mechanism-based treatment strategies. These strategies may focus on modulation of GABAergic mechanisms or the prevention of neuronal damage, cell stress, and cell destruction, or even the rebuilding of functional synapses.

BASIS OF NEW STRATEGIES FOR THERAPEUTIC INTERVENTION

The challenge of finding new and more effective antiepileptic pharmacotherapeutics, particularly drugs that will control difficult to treat and refractory epileptic states, will require the integration of data obtained by new methodologies with new insights into understanding neurotransmission in convulsive disorders. The focus of this chapter on human epilepsy and the role of GABA necessitates omitting the detailed discussion of excitatory neurotransmitters and their receptors, even though they are likely to play a role and therefore represent valid targets for anticonvulsant agents. Instead the goal is to bring together evidence involving a common denominator to both excitatory amino acid and GABA inhibitory mechanisms, i.e., ionic homeostasis, particularly that of Ca^{2+}.

It is clear that there are many avenues for enhancing GABA-mediated inhibitory processes in animal models of epilepsy. Many, such as increasing GABA affinity at the GABA receptor-benzodiazepine receptor complex or prolonging the open state of GABA mediated chloride conductance channel as produced by barbiturates, have been discussed in other chapters. Similarly, agents that inhibit GABA aminotransferase and agents that interfere with re-uptake systems are reviewed elsewhere in this book. A few of these approaches, because of a variety of reasons, usually toxicity, are unlikely treatment modalities. GABA agonists such as muscimol and THIP fall into this category. Alternatively, pharmacokinetic repackaging into liposomes, enabling the polar GABA to gain direct access to the CNS, is being investigated (Mori et al., 1989; Dichter, 1989) but thus far little information is available on the clinical effectiveness. An indirect way to enhance the effectiveness of inhibitory GABA mechanisms is blocking *N*-methyl-D-aspartate receptors with drugs such as MK-801. Clinical trials have proved less efficacious than anticipated though, mainly because of side effects and pharmacokinetic difficulties in maintaining therapeutic levels of the drug (see Dichter, 1989).

Diabetic patients who undergo hypoglycemic episodes have an increased tendency to develop seizures. Amoroso et al. (1990) have suggested that decreased blood glucose will activate substantia nigra ATP-regulated potassium (K_{ATP}) channels leading to hyperpolarization and a consequent decrease in GABA release. These K_{ATP} channels are regulated by intracellular ATP:ADP ratios and when ATP levels decrease, they open and produce a hyperpolarization. They also exhibit high affinity binding for sulfonylureas, an important class of antidiabetic drugs. In rat substantia nigra slices for instance, these drugs give rise to GABA release with EC_{50} values from 3 nM (gliquidone) to 300 µM (tolbutamide). Presumably this effect is the result of binding to the K_{ATP} channels directly, causing a depolarization and activation of L-type Ca^{2+} channels, a rise in intracellular Ca^{2+} and a subsequent neurotransmitter release. Therefore, any condition which decreases ATP levels, anoxia and ischemia, could among other changes be expected to decrease the inhibitory effects of GABA by opening K_{ATP} channels, leading to hyperpolarization. Similarly, any drug that can interfere with K_{ATP} channel function directly, or indirectly by altering the cellular energy charge, has the potential for antiepileptic activity.

Another provocative new insight into neuronal regulation involves the presence of an ATP-dependent chloride transport system in rat brain synaptosomes. Shiroya et al. (1989) describe an presumably outwards directed chloride pump which was half maximally inhibited by 57 µM ethacrynic acid, partially inhibited by 1 mM uanadate, and completely inhibited by 100 µM N-ethylmaleimide. Chloride uptake into these vesicles was unaffected by ouabain (1mM) and oligomycin (1µM). Uptake was shown to be susceptible to both pH and temperature with rather narrow optima of 7.4 and 37–42°C, respectively. Any deviation from these parameters, as for example in febrile states, could be expected to significantly alter the neuronal homeostasis, resulting in a less stable membrane potential. Conversely, an increase in pump-ATPase activity and an increase in the Cl^- gradient would result in the stabilization of the membrane potential. An alteration in other Cl^- flux systems such as stimulation of Cl^- co-transport carriers would have the same result.

So far this discussion has centered on the regulation of ionic species that are primarily involved in the initial or instantaneous stimulus response or short-term electrophysiological control of the neuronal cell. Another important aspect of epilepsy, particularly as it relates to etiology, prevention of seizures, pathology and treatment, is the well known but poorly understood long-term effects of prolonged and repetitive seizure activity. Ca^{2+} is intimately involved in both short-term as well as long-term cellular regulation, and under pathological conditions may be one of the main factors contributing to cellular toxicity and eventual cell death. Not only does the cation appear to be essential in neurotransmitter release but it plays a crucial role in signal transduction, second messenger and nuclear anabolic mechanisms.

Presynaptic neurons have a number of receptor-activated and voltage-sensitive Ca^{2+} channels, all of which may have a role in epileptic activity

(Greenberg, 1987). Rather than excessively active Ca^{2+} currents, a relative lack or diminished functioning of opposing, yet often coupled Ca^{2+}-activated potassium channels may play a part in epileptic events. Analogous to other well studied stimulus-release coupling events in the peripheral nervous system, Asakura et al. (1982) demonstrated that extracellular Ca^{2+} needs to be present during depolarization for GABA release to occur. In Ca^{2+} ionophore A23187-treated synaptosomes, GABA is released in a Ca^{2+}-dependent manner over a concentration range of < 1 µM to about 3 mM, with a half maximal activation occurring at 30 µM $Ca^{2+}{}_o$. Interestingly, stripping of synaptosomal membranes of Ca^{2+} by addition of Ca^{2+} chelators also leads to specific GABA release from superfused mouse brain synaptosomes. This effect is Na^+-dependent and prevented in the presence of Mg^{2+}, and thought to be due to membrane destabilization from the lack of membrane Ca^{2+} association (Arias et al., 1984). Carvalho et al. (1986) demonstrated high affinity binding of several various types of Ca^{2+} channel entry blockers to synaptosomal membranes. These same drugs also inhibited Ca^{2+}-dependent and Na^+-dependent (Ca^{2+}-independent) GABA release, albeit at considerably higher doses. There was a correlation of inhibition of Ca^{2+} uptake and inhibition of GABA release by verapamil. Each of these effects exhibited an IC_{50} of > 10 µM which represents a large discrepancy from the value for high affinity binding of verapamil to the channel. The authors concluded that the effects of these drugs on Ca^{2+} fluxes and GABA release are not mediated through the high affinity binding sites on the synaptosomal membranes. Similar findings have also been reported for other neuronal transmitter release (Karaki et al., 1984) and other membrane Ca^{2+} transport phenomena in human erythrocyte model systems (Raess and Record, 1990). More directly related to epilepsy is the fact that verapamil in experimental animals was very effective in inhibiting epileptiform discharges in a penicillin focus (Walden et al., 1985). A possible explanation may be that verapamil has been shown to inhibit the GABA removal process at the synaptic cleft (Liron et al., 1985; King and Tunnicliff, 1990). In addition, flunarizine, another type of Ca^{2+} antagonist, was effective as an adjunct drug in a placebo controlled crossover clinical trial, resulting in significantly reduced seizure activity in patients not adequately controlled by other treatments (Vanden Bussche et al., 1986).

Postsynaptically, Ca^{2+} has been reported to decrease the number of $GABA_A$ sites while increasing $GABA_B$ sites. While it is difficult to extrapolate these in vitro observations to epileptic events in patients, it has been offered as a plausible explanation for the often observed desensitization of GABA inhibitory effects due to excessive cell firing (Corda and Guidotti, 1983).

Ca^{2+} has also been implicated as the complication in postictal cellular stress and other, often irreversible, cellular changes including nuclear DNA genomic changes. There are several cellular safeguards against a damaging cellular Ca^{2+} overload, i.e., Ca^{2+} extrusion mechanisms such as the Na^+-Ca^{2+} exchanger and the Ca^{2+}-pump, intracellular organelle Ca^{2+} sequestering mechanisms such as ER and mitochondria, and Ca^{2+} buffering proteins such as calmodulin. Despite these control mechanisms to keep intracellular Ca^{2+} within low limits, it is conceivable that during generalized convulsive seizure activity precipitating dramatic drops of extracellular Ca^{2+} (Pumain et al., 1985) with concomitant rises in intracellular Ca^{2+} (Connor et al., 1988), the capacity of such systems is simply overtaxed. Cell damage and eventual cell death by Ca^{2+} overload due to prolonged seizure activity (Olney et al., 1986) may actually offer an explanation for structural damages subsequent to *status epilepticus*, prolonged febrile seizure activity in children (Sagar and Oxbury, 1987), and other forms of epilepsy (Mouritzen-Dam, 1980; Babb et al., 1984; 1989). Not all increases in cellular Ca^{2+} need to be detrimental and cause irreversible cellular damage. Tightly controlled cellular fluctuations are a normal part of cellular signal transduction including those of GABAergic mechanisms, but if exaggerated as can be shown in experimentally induced seizure activity, Ca^{2+} can effect neural plasticity via cytosolic and nuclear second messenger systems. Of particular interest for seizure induced long-term changes is the Ca^{2+}-induced cascade of cellular and nuclear events that leads to the expression of c-*fos* proto-oncogene, mRNA and the resulting gene product nuclear c-*fos* protein, which is thought to be involved in cell growth and differentiation (Dragunow and Robertson 1987; 1988; Gloor, 1989).

In absence seizures, GABAergic inhibitory influences appear to be pre-served and consequently N-methyl-D-aspartate-induced postsynaptic Ca^{2+} stress is less likely to occur (Gloor, 1989). In other types of epilepsy, two main therapeutic objectives should be the short-term control of initiation and spreading of epileptogenic discharge activity, and perhaps even more import-ant, the prevention of semi-permanent or permanent neuronal damage. With many potential targets for drug modulation of inhibitory or excitatory syn-apses, the control of Ca^{2+} homeostasis, pre- and postsynaptically, appears to emerge as a productive arena for the development of effective antiepileptics.

CONCLUSIONS

Although there is no conclusive evidence implicating GABA per se in various forms of human epilepsy, it is hoped that with the advent of newer, more sophisticated techniques and more rigorous experimental design, some of the apparent ambiguities will be resolved. Considering the ubiqui-tous nature of GABA regulated pathways, various ionic conductances and their regulation at pre- and postsynaptic sites, and the multitude of relevant receptors underlying excitatory and inhibitory neurotransmitters, it is not surprising that our understanding of epileptogenic activity, especially in humans, is far from complete. Most of the drugs used today to treat epilepsy are of the "shotgun" variety, yet a number of specific agents for specific disorders which can prevent seizure activity or their harmful consequences are needed. It is not likely that any given class of drugs, acting via a single mechanism of action, will be useful in achieving this goal. It is obvious that there is a need to add to the armamentarium by integrating several new lines of evidence in order to achieve a rational basis for designing new and more effective antiepileptic drugs.

REFERENCES

Amoroso S, Schmid-Antomarchi H, Fosset M, Lazdunski M: Glucose, sulfonylureas, and neurotransmitter release: Role of ATP-sensitive K^+ channels. Science 247: 852–854, 1990.

Araki K, Harada M, Ueda Y, Takino T Kuriyama K: Alteration of amino acid content of cerebrospinal fluid from patients with epilepsy. Acta Neurol Scand 78: 473–479, 1988.

Arias C, Sitges M, Tapia R: Stimulation of [^{3}H]γ-aminobutyric acid release by calcium chelators in synaptosomes. J Neurochem 42: 1507–1514, 1984.

Asakura T, Hoshino M, Kobayashi T: Effect of calcium ion on the release of γ-aminobutyric acid from synaptosomal fraction. J Biochem (Tokyo) 92: 1919–1923, 1982.

Babb TL, Brown WJ, Pretorius JK, Davenport CJ, Lieb JP, Crandall PH: Temporal lobe volumetric cell densities in temporal lobe epilepsy. Epilepsia 25: 729–740, 1984.

Babb TL, Pretorius JK, Kupfer WR, and Crandall PH: Glutamate decarboxylase-immunoreactive neurons are preserved in human epileptic hippocampus. J Neurosci 9: 2562–2574, 1989.

Bakay RAE, Harris AB: Neurotransmitter receptor and biochemical changes in monkey cortical epileptic foci. Brain Res 206: 387–404, 1981.

Bartenstein P, Ludolph A, Schober O, Lottes G, Böttger I, Beer H-F: Vergleich von Blutfluss und Benzodiazepin-Rezeptorverteilung bei fokaler Epilepsie: Vorläufige Ergebnisse einer SPECT-Studie. Nucl -Med 28: 181–186, 1989.

Boehme DH, Cottrell JC, Leonberg SC, Zeman W: A dominant form of neuronal ceroid-lipofuscinosis. Brain 94: 745–760, 1971.

Bohlen P, Huot S, Palfreyman MG: The relationship between GABA concentration in brain and cerebrospinal fluid. Brain Res 167: 297–305, 1979.

Braak H, Goebel HH: Loss of pigment-laden stellate cells: A severe alteration of the isocortex in juvenile neuronal ceroid-lipofuscinosis. Acta Neuropathol (Berl) 42: 53–57, 1978.

Carvalho CM, Santos SV, Carvalho AP: γ-Aminobutyric acid release from synaptosomes as influenced by Ca^{2+} and Ca^{2+} channel blockers. Eur J Pharmacol 131: 1–12, 1986.

Connor JA, Wadman WJ, Hockberger PE, Wong RKS: Sustained dendritic gradients of Ca^{2+} induced by excitatory amino acids in CA1 hippocampal neurons. Science 240: 649–653, 1988.

Corda MG, Guidotti A: Modulation of GABA receptor binding by Ca^{2+}. J Neurochem 41: 277–280, 1983.

Crawford PM, Chadwick DW: GABA and amino acid concentrations in lumbar CSF in patients with treated and untreated epilepsy. Epilepsy Res 1: 328–338, 1987.

Dichter MA: Cellular mechanisms of epilepsy and potential new treatment strategies. Epilepsia 30 (Suppl. 1): S3–S12, 1989.

Dragunow M, Robertson HA: Kindling stimulation induces c-*fos* protein(s) in granule cells of the rat dentate gyrus. Nature 329: 441–442, 1987.

Enna SJ, Ziegler MG, Lake CR, Wood JH, Brooks BR, Butler IJ: Cerebrospinal fluid γ-aminobutyric acid: Correlation with cerebrospinal fluid and blood constituents and alterations in neurological disorders. In Wood JH (ed): "Neurobiology of Cerebrospinal Fluid I." New York: Plenum, 1980.

Enna SJ, Wood JH, Snyder SH: γ-Aminobutyric acid (GABA) in human cerebrospinal fluid: Radioreceptor assay. J Neurochem 28: 1121–1124, 1977.

Farde L, Hall H, Ehrin E, Sedvall G: Quantitative analysis of D_2 dopamine receptor binding in the living human brain by PET. Science 231: 258–261, 1986.

Feldblum S, Ackermann RF, Tobin AJ: Long-term increase in glutamate decarboxylase mRNA in a rat model of temporal lobe epilepsy. Neuron 5: 361–371, 1990.

Gloor P: Epilepsy: Relationship between electrophysiology and intracellular mechanisms involving second messengers and gene expression. Can J Neurol Sci 16: 8–21, 1989.

Greenberg D: Calcium channels and calcium channel antagonists. Ann Neurol 21: 317–330, 1987.

Grossman MH, Hare TA, Manyam NVB Glaeser BS, Wood JH: Stability of GABA levels in CSF under various conditions of storage. Brain Res 182: 99–106, 1980.

Grove J, Schechter PJ, Hsanke NFJ, de Smet Y, Agid Y, Tell G, Koch-Weser J: Concentration gradients of free and total γ-aminobutyric acid and homocarnosine in human CSF: Comparison of suboccipital and lumbar sampling. J Neurochem 39: 1618–1622, 1982.

Hare TA, Manyam NVB, Glaeser BS: Evaluation of cerebrospinal fluid γ-aminobutyric acid content in neurologic and psychiatric disorders. In Wood JH (ed): "Neurobiology of Cerebrospinal Fluid I." New York: Plenum, 1980.

Houser CR, Harris BA, Vaughn JE: Time course of the reduction of GABA terminals in a model of focal epilepsy: A glutamic acid decarboxylase immunocytochemical study. Brain Res 383: 129–145, 1986.

Karaki H, Nakagawa H, Urakawa N: Effects of calcium antagonists on release of [^{3}H]noradrenaline in rabbit aorta. Eur J Pharmacol 101: 177–183, 1984.

Killam KF, Bain JA: Convulsant hydrazides I. In vitro and in vivo inhibition of vitamin B_6 enzymes by convulsant hydrazides. J Pharmacol Exp Ther 119: 255–262, 1957.

King SM, Tunnicliff G: Na^+ and Cl^- dependent [^{3}H]GABA binding to catfish brain particles. Biochem Int 20: 821–831, 1990.

Knight M, Ebert J, Parish RA, Berry H, Fogelson H: γ-Aminobutyric acid in CSF of children with febrile convulsions. Arch Neurol 42: 474–475, 1985.

Kuroda H, Ogawa N, Yamawaki Y, Nukina I, Ofuji T, Yamamoto M, Otsuki S: Cerebrospinal fluid GABA levels in various neurological and psychiatric diseases. J Neurol Neurosurg Psychiat 45: 257–260, 1982.

Liron Z, Roberts E, Wong E: Verapamil is a competitive inhibitor of γ-aminobutyric acid and calcium uptakes by mouse brain subcellular particles. Life Sci 36: 321–327, 1985.

Lloyd KG, Bossi L, Morselli PL, Munari C, Rougier M, Loiseau H: Alterations of GABA-mediated synaptic transmission in human epilepsy. Adv Neurol 44: 1033–1044, 1986.

Lloyd KG, Munari C, Bossi L, Stoeffels C, Talairach J, Morselli PL: Biochemical evidence for the alterations of GABA-mediated synaptic transmission in pathological brain tissue (stereo EEG or morphological definition) from epileptic patients. In Morselli PL, Lloyd KG, Löscher W, Meldrum B, and Reynolds EH (eds): "Neurotransmitters, Seizures, and Epilepsy." New York: Raven Press, 1981, pp. 325–338.

Lloyd KG, Munari C, Bossi L, Morsielli PL: Neurochemical evidence for the GABA hypothesis of human epilepsy. In Porter RJ, Mattson RH, Ward AA, Dam M (eds): "Advances in Epileptology: XVth Epilepsy International Symposium" New York: Raven Press, 1984, pp.3–7.

Löscher W, Rating D, Siemes H: GABA in cerebrospinal fluid in children with febrile convulsions. Epilepsia 22: 679–702, 1981.

Marksteiner J, Sperk G: Concomitant increase of somatostatin, neuropeptide Y, and glutamate decarboxylase in frontal cortex of rats with decreased seizure threshold. Neuroscience 26: 379–385, 1988.

McGeer PL, McGeer EG, Wada JA: Glutamic acid decarboxylase in Parkinson's disease and epilepsy. Neurology 21: 1000–1007, 1971.

Mori N, Terayama K, Sato T, Kumashiro H: [Suppressive effect of GABA-containing liposomes on kindled convulsion.] No To Shinkei 41: 193–198, 1989.

Mouritzen-Dam A: Epilepsy and neuron loss in the hippocampus. Epilepsia 21: 617–629, 1980.

Olney JW, Collins RC, Sloviter RS: Excitotoxic mechanisms of epileptic brain damage. Adv Neurol 44: 857–877, 1986.

Paul SM, Skolnick P: Rapid changes in brain benzodiazepine receptors after experimental seizures. Science 202: 892–893, 1978.

Perry TL, Hansen S, Kennedy J, Wada JA, Thompson G: Amino acids in human epileptogenic foci. Arch Neurol 32: 752–754, 1975.

Pitkanen A, Matilainen R, Halonen T, Kutvonen R, Hartikainen P Riekkinen P: Inhibitory and excitatory amino acids in cerebrospinal fluid of chronic epileptic patients. J Neural Transm Suppl 76: 221–230, 1989.

Pumain R, Ménini C, Heinemann U, Louvel J, Siva-Barrat, C: Chemical synaptic transmission is not necessary for epileptic seizures to persist in the baboon Papio papio. Exp Neurol 89: 250–258, 1985.

Raess BU, Record DM: Inhibition of erythrocyte Ca^{2+} pump by Ca^{2+} antagonists. Biochem Pharmacol 40: 2549–2555, 1990.

Ribak CE, Harris AB, Vaughn JE, Roberts E: Inhibitory, GABAergic nerve terminals decrease at sites of focal epilepsy. Science 205: 211–214, 1979.

Rougier A, Loiseau H, Cohadon F, Lloyd KG, Morselli PL: GABA and other amino acids in cortical and hippocampal foci removed neurosurgically from epileptic patients. In Porter RJ, Mattson RH, Ward AA, and Dam M (eds): "Advances in Epileptology: XVth Epilepsy International Symposium." New York: Raven Press, 1984, pp. 43–48.

Sagar HJ, Oxbury JM: Hippocampal neuronal loss in temporal lobe epilepsy: Correlation with early childhood convulsion. Ann Neurol 22: 334–340, 1987.

Savic I, Roland P, Sedvall G, Perrson A, Pauli S, Widen L: In vivo demonstration of reduced benzodiazepine receptor binding in human epileptic foci. Lancet 2 (8616): 863–866, 1988.

Schmidt D, Cornaggia C, Löscher W: Comparative studies of the GABA system in neurosurgical brain specimens of epileptic and non-epileptic patients. In Fariello RG, Morselli PL, Lloyd KG, Quesney LF, Engel J (eds): "Neurotransmitters, Seizures and Epilepsy, II." New York: Raven Press, 1984, pp. 275–283.

Sherwin AL, Quesney F, Gauthier S, Olivier A, Robitaille Y, McQuaid P, Harvey C, Van Gelder N: Enzyme changes in actively spiking areas of human epileptic cerebral cortex. Neurology 34: 927–933, 1984.

Sherwin A, Robitaille Y, Quesney F, Olivier A, Villemure J, Leblanc R, Feindel W, Andermann E, Gotman J, Andermann F, Ethier R, Kish S: Excitatory amino acids are elevated in human epileptic cerebral cortex. Neurology 38: 920–923, 1988.

Sherwin Al, Mathew E, Blair M, Gué D: Benzodiazepine receptor binding is not altered in human epileptic cortical foci. Neurology 36: 1380–1382, 1986.

Shiroya T, Fukunaga R, Akashi K, Shimada N, Takagi Y, Nishino T, Hara M, Inagaki C: An ATP-driven Cl⁻ pump in the brain. J Biol Chem 264: 17416–17421, 1989.

Tunnicliff G: Action of inhibitors on brain glutamate decarboxylase. Int J Biochem 22: 1235–1241, 1990.

Vanden Bussche G, Wauquier A, Ashton D, deBeukelaar F: In Meldrum B, Porter R (eds): "New Anticonvulsant Drugs." London: John Libbey Press, 1986.

Van Gelder NM, Courtois A: Close correlation between changing content of specific amino acids in epileptogenic cortex of cats, and the severity of epilepsy. Brain Res 43: 477–484, 1972.

Van Gelder NM, Sherwin AL, Rasmussen T: Amino acid content of epileptogenic human brain: Focal versus surrounding regions. Brain Res 40: 385–393, 1972.

Walden J, Speckman E-J, Witte DW: Suppression of focal epileptiform discharges by intraventricular perfusion of a calcium antagonist. Electroencephalogy Clin Neurophysiol 61: 299–309, 1985.

Wood JH, Hare TA, Glaeser BS, Ballenger JC, Post RM: Low cerebrospinal fluid γ-aminobutyric acid content in seizure patients. Neurology 29: 1203–1208, 1979.

Zeman W: The neuronal ceroid-lipofuscinoses-Batten-Vogt syndrome: A model for human aging? Adv Gerontol Res 3: 147–170, 1971.

GABA Mechanisms in Epilepsy, pages 121–147
© 1991 Wiley-Liss, Inc.

6
GABA Dysfunction in Animal Models of Epilepsy

ROGER W. HORTON

Department of Pharmacology and Clinical Pharmacology,
St. George's Hospital Medical School, London SW17 ORE, England

INTRODUCTION

Improved treatment of the epilepsies is dependent upon gaining a better understanding, at the cellular and molecular levels, of the pathophysiological processes that underlie convulsive disorders. Practical and ethical difficulties limit the extent to which such processes can be studied in man; the study of models of epilepsy in experimental animals provides an alternative approach. A large number of animal models of epilepsy have been developed; they are characterised by their uniformly high susceptibility to seizures, either occurring spontaneously or induced by sensory stimulation. The range of species is diverse from mice to sub-human primates. Each model has its characteristic features in terms of seizure onset, severity, chronicity, and EEG correlates. These models, although imperfect replicas of human epilepsies, provide the opportunities to study neurochemical events that predispose to seizures, that precede seizures, and that are the consequences of single or multiple seizures.

There is overwhelming evidence that γ-aminobutyric acid (GABA) is a major inhibitory neurotransmitter in the mammalian CNS. Pharmacological manipulation of GABA-mediated neurotransmission has profound effects upon seizure susceptibility in normal laboratory animals; a reduction in GABA function is associated with lowered seizure threshold and spontaneous convulsions, whereas enhancing GABA-mediated inhibition is often associated with an increase in seizure threshold. Many studies have investigated the involvement of altered GABA-mediated inhibition in animal models of epilepsy. The aim of this chapter is to review this evidence and to integrate neurochemical and other data. In an attempt to provide a balanced view of the relative importance of GABA, evidence for the involvement of other neurotransmitters is also summarised. General metabolic, endocrine or other abnormalities not directly related to specific neurotransmitter systems, although of possible importance in some animal models of epilepsy, are not considered.

Emphasis has been placed on widely studied in vivo models of a semi-chronic or chronic nature. Acute models involving systemic or focal administration of convulsant drugs or electrical stimulation and in vitro models are not considered. Animal models of epilepsy can be arbitrarily classified into those which are naturally occurring or have been developed by selective breeding (genetic models) and those induced physically or chemically (inductive models).

GENETIC MODELS OF EPILEPSY
Genetically Epilepsy Prone Rats (GEPR)

Background and phenomenology. Selective breeding of rats has given rise to colonies with high susceptibility to sound induced seizures. Originally termed audiogenic seizure susceptible (AGS) rats, their development into the present GEPR colonies and associated nomenclature has been reviewed by Reigel et al. (1986).

Intense auditory stimulation of the GEPR produces one or more periods of wild running which progress through clonic phases to a tonic extension, which is rarely fatal. The descriptive scoring scale of Jobe (1981) is widely used as an index of seizure severity. Audiogenic seizures can be induced by 3 weeks of age and susceptibility does not decline with age. The running and clonic phases are not accompanied by distinct cortical epileptiform EEG activity and are probably of brain stem origin; seizures only projecting to the cortex during repeated tonic seizures (Tacke et al., 1984). Lesion studies also indicate that subcortical structures are primarily important in the manifestation of audiogenic seizures; lesions of the inferior colliculus (IC), the ventral cochlear nuclei and the pontine reticular formation but not the auditory cortex prevent seizures (see Laird and Jobe, 1987). Although audiogenic seizures have been most widely studied, GEPRs have increased sensitivity to other seizure provoking procedures, such as hyperthermia, electroshock and certain chemical convulsants (Reigel et al., 1986).

GABA in the GEPR. Several lines of evidence suggest that abnormalities of GABA function in the IC may be responsible for audiogenic seizure susceptibility in the GEPR. Bilateral injection of the $GABA_A$ agonist, muscimol, or the GABA-transaminase inhibitor, gabaculline, into the IC of the GEPR blocks seizure susceptibility, suggesting a possible impairment of GABA function (Faingold, 1988). Direct application of GABA or benzodiazepines (BZ) inhibits the firing of IC neurons. However, the iontophoretic current required to produce comparable inhibition in the GEPR is twice that required in control Sprague-Dawley (SD) rats (Faingold et al., 1986), suggesting a reduced post-synaptic efficacy of GABA in the IC of the GEPR.

Following intense acoustic stimulation, a post stimulus afterdischarge is often seen in IC neurons. In the presence of bicuculline, the after discharge is enhanced suggesting that GABA-mediated mechanisms normally suppress this discharge. In the GEPR the iontophoretic current of bicuculline required to produce the post-stimulus after discharge in the IC is significantly lower than in SD rats, again suggesting a reduced efficacy of GABA-mediated inhibition (Faingold, 1988).

Anatomical studies of the IC of the GEPR indicate an increase in the total number of neurons, and perhaps somewhat surprisingly an increase in the number of neurons that stain positively with an antibody directed against glutamic acid decarboxylase (GAD), compared to non-epileptic SD rats (Roberts et al., 1985b). Twenty five percent of neurons stained positively for GAD in the IC of SD rats, whereas the proportion was 35% in the GEPR. The higher number of GAD-positive neurons was most marked in the central nucleus of the IC and was due to a selective increase of 200% in the small cell body (10–15 micron diameter) neurons and of 90% in the medium cell body (15–25 micron diameter) neurons. The number of total neurons and GAD-positive neurons did not differ between the GEPR and the SD rat in the medial superior olive, another primary auditory brain stem nucleus or in the oculomotor nucleus, a brain stem nucleus not related to the primary auditory pathway, suggesting that the differences may be restricted to the IC.

The increase in neuronal cell number in the IC appears not to be a compensatory change related to seizure activity because young GEPRs, who had not had seizures, also showed the increased numbers, at least for the small neurons (Roberts et al., 1985a). The number of medium sized neurons was lower in the GEPR than the SD rat at 4 days of age, similar in number at 10 days of age and increased in adults.

GABA concentration in the central nucleus of the IC, the area showing the most marked increase in the number of GAD-positive neurons, has been reported to be 2.3 times higher in the GEPR than the SD rat (Ribak et al., 1988). Other studies of the whole IC have reported GABA concentrations that do not differ (Huxtable et al., 1982), or are 30-40% higher in the GEPR than in the SD rat (Chapman et al., 1986). GAD activity measured enzymically, in the presence or absence of exogenous pyridoxal phosphate, did not differ in IC, cortex or striatum between GEPR and controls (Huxtable et al., 1982).

Booker et al. (1986) measured GABA receptors (labeled with [^{3}H] muscimol) in the cerebral cortex of the GEPR. In a stable colony that developed only mild myoclonic responses to auditory stimulation (GEPR-3) no differences in GABA binding were found compared to controls. The number of high and low affinity GABA$_A$ binding sites was, however, higher than controls in a stable colony displaying severe seizures (GEPR-9) in response to auditory stimulation. BZ binding was also higher in the GEPR-9 compared to controls, although the difference did not reach statistical significance (Booker et al., 1986). Tacke and Braestrup (1984) have studied various aspects of GABA receptor binding in the GEPR. The binding of a BZ agonist, a BZ inverse agonist and the chloride channel ligand [^{35}S]TBPS did not differ from controls in cerebral cortex, hippocampus, brain stem or cerebellum. However, muscimol enhancement of [^{3}H]diazepam binding, a putative index of the GABA recognition site-BZ coupling, was significantly lower in the hippocampus but not other brain regions of the GEPR. Muscimol enhancement of BZ binding did not, however, differ in the hippocampus of the Krushinsky-Molodkina strain of AGS rats compared to controls. The above studies are difficult to interpret because the GEPR had previously undergone convulsions. Mimaki et al. (1984) studied BZ binding in whole brain membranes from GEPR-9s that had not undergone convulsions and found an increased number of sites compared to controls.

Other neurotransmitters in GEPR. Abnormalities of excitatory amino acid (EAA) - mediated neurotransmission have been reported in the GEPR. Chapman et al. (1986) found aspartate concentrations to be 30–40% lower in the hippocampus, striatum and IC, and glutamate concentrations 20–30% lower in hippocampus and striatum of non-convulsed GEPRs compared to controls. Following a seizure, aspartate concentration in the IC was markedly increased. Enhanced K^+-stimulated release of preloaded $[^3H]$-glutamate has been reported from the hippocampus of the GEPR in vivo (Lehman, et al., 1986), although not in vitro (Chapman et al., 1986). The number of $[^3H]$-glutamate binding sites has been reported to be higher in the hippocampus of the GEPR than controls (Mills et al., 1985). Although these data are suggestive of abnormal EAA function in the hippocampus of the GEPR, focal injection of the EAA antagonist 2-amino-5-phosphopentanoic acid (AP5) into this structure has little effect on audiogenic seizures in the GEPR, whereas much smaller doses of AP5 into the IC blocks audiogenic seizures (Faingold, 1988; Faingold et al., 1988).

Of the monoamine neurotransmitters, noradrenaline (NA) has been most extensively studied in the GEPR. NA concentration, NA turnover, and tyrosine hydroxylase activity are lower in many brain areas of non-convulsed GEPRs (Dailey and Jobe, 1986; Dailey et al., 1989). NA uptake into synaptosomes prepared from IC, cortex, amygdala, and hippocampus of GEPR-9 was lower than in non-epileptic controls (Browning et al., 1986). In cortex, uptake velocity was reduced while affinity was unaltered, indicating a reduction in the number of uptake sites. Taken together these data are consistent with a reduced NA innervation in GEPRs compared to control rats. Although some studies of adrenergic binding sites have found no differences (Ko et al., 1984; Booker et al., 1986), a significantly lower number of α_1-adrenoceptors and reduced NA-stimulated inositol phospholipid turnover in the frontal cortex, but not the hippocampus, of GEPR-9s has been reported (Nicoletti et al., 1986).

Epileptic Gerbils

Background and phenomenology. Mongolian gerbils develop EEG and motor seizures in response to a variety of stimuli, such as photic or audiogenic stimulation, handling, or simply by being placed in a novel environment. The wide range of provoking stimuli and the variability in response have tended to limit studies in this model, although these problems can be overcome with selective breeding and standardization of the precipitating stimulus. Development of seizure susceptibility appears to be similar in randomly bred and selectively bred strains. The age of onset is about 6–8 weeks, with maximum sensitivity at about 7 months, which is then maintained.

Seizures can be evoked in more than 98% of randomly bred gerbils by a blast of compressed air (5 bar for 10 sec) directed at the animal's back, whereas less than 10% developed seizures in response to handling, photic or auditory stimulation or a change in environment (Löscher, 1984). With the use of this technique the effects of clinically used antiepileptic drugs have been studied in the gerbil (Löscher, 1984). Benzodiazepines, valproate, and ethosuximide were effective against minor (myoclonic) seizures, whereas phenytoin was ineffective and carbamezepine only partially effective. Phenytoin, phenobarbitone, primidone, and carbamazepine were effective against major (mostly generalised tonic-clonic) seizures.

GABA in epileptic gerbils.　A range of experimental drugs have also been studied against major seizures in the gerbil. Direct and indirect GABA-mimetics were more active than most clinically used antiepileptic drugs (except benzodiazepines) and were also more active in the gerbil than in the pentylenetetrazol and electroshock models and audiogenic seizures in DBA/2 mice (Löscher et al., 1983). Seizures in gerbils were not sensitive to alterations in acetylcholine, noradrenaline, serotonin, glycine, or excitatory amino acid mediated neurotransmission (Löscher, 1985). These results suggest that seizure susceptibility in the gerbil may be critically related to some defect in GABA-mediated neurotransmission.

Selective breeding of gerbils has produced seizure susceptible (SS) and seizure resistant (SR) colonies of gerbils. The concentration of GABA is lower in the parietal and temporal lobes and thalamus of SS than SR gerbils, although the difference was only statistically significant for the parietal lobe (Lomax et al., 1987). Olsen et al., (1984; 1986) have compared GABA and BZ binding in SS and SR strains. No significant difference were found in GABA binding in cerebral cortex, cerebellum, or thalamus/midbrain. There was a tendency for GABA binding to be lower in pons-medulla of SS than SR gerbils, although this did not reach statistical significance. BZ binding was significantly lower in thalamus/mid brain and caudal midbrain of SS than in SR gerbils. Further autoradiographic analysis localized the deficits in BZ binding to the substantia nigra and the periaquaductal gray region. The lower BZ binding in the substantia nigra may be of functional significance, since this structure has been shown to be important in controlling the motor expression of seizures (Iadarola and Gale, 1982; Garant and Gale, 1983; McNamara et al., 1983; Millan et al., 1988).

Although BZ binding did not differ between SS and SR gerbils in the hippocampus, other differences have been reported in this structure. A loss of spines on hippocampal pyramidal cell dendrites has been reported, as has an increased number of GAD-positive neurons and terminals in the dentate gyrus (Peterson et al., 1985). Many GAD positive cells in the dentate gyrus have been shown to be basket cells, which mediate feedback inhibition of the excitatory granule cells. However, the anatomical appearance of the granule cells suggests that they are more rather than less active in the SS gerbil. The situation appears analogous to the IC in the GEPR. In both situations increased numbers of GAD-positive neurons are associated with an increase in excitability.

Other neurotransmitters in epileptic gerbils. Although other neurotransmitter systems have been less extensively studied, there is pharmacological and neurochemical evidence suggesting an involvement of opioid mechanisms. Opioid agonists and β-endorphin produce a naloxone-sensitive decrease in seizures (Lee et al., 1984). Autoradiographic quantitation of opiate receptors (using [^{3}H]dihydromorphine at a concentration which labels largely μ-receptors) demonstrated significantly higher binding in SS compared to SR gerbils, particularly in the substantia nigra (by 98%), periaqueductal gray (by 91%) and medial geniculate body (by 62%) (Lee et al., 1986). Significantly higher binding was also seen in the superior colliculus, interpeduncular nucleus, accessory optic nuclei, hippocampus and dentate gyrus of the SS gerbil, although the magnitude of the differences was less in these structures (25–51%). However, since the SS gerbils used in this study had long histories of seizures, it is unclear whether the differences in opiate binding are responsible for, or contribute to the mechanisms of seizure susceptibility or whether they are an adaptive response to repeated seizures.

El Mice

Background and phenomenology. The El mouse is an inbred mutant strain derived from ddY mice, that develops tonic-clonic seizures in response to vestibular stimulation (tossing in the air, horizontal swinging, alternating rotation). Weekly vestibular stimulation starting at 4–5 weeks of age results in animals that by 10–15 weeks regularly convulse on appropriate stimulation, a sensitivity that then persists throughout life. Adult mice that have not received vestibular stimulation while young also sometimes develop seizures spontaneously or in response to vestibular stimulation. For assessing neurochemical data El mice that have received vestibular stimulation (seizure-prone) are most often compared to El mice that have not received stimulation or to non-seizure susceptible mice strains (gpc, ddY). Using 2-deoxyglucose autoradiography, Suzuki et al. (1983) showed that the seizures in El mice originate in the hippocampus or deep temporal lobe and then spread to other brain regions. A variety of anatomical differences in the hippocampus have been described when El and ddY mice are compared; particularly atrophic changes in pyramidal cell somata.

GABA in El mice. Hattori et al. (1985) measured GABA and BZ binding in membranes prepared from whole brain of 22–24 week old El and ddY mice. In El mice that had received vestibular stimulation at weekly intervals from 4 weeks of age and had experienced a mean of 14 seizures, the number of high affinity [3H] muscimol binding sites was higher by 31% than in ddY mice, although binding affinity was lower in El mice (K_D 1.96 v 1.46nM). It seems unlikely that the difference in [3H]muscimol binding was related to the seizures, because binding was similar in El mice that had not received vestibular stimulation or developed convulsions. BZ binding did not differ between ddY and El mice that had or had not developed convulsions. In contrast to a previous report of increased GABA concentrations in El mice (Naruse et al., 1960), GABA concentrations did not differ between ddY mice and stimulated or unstimulated El mice (Hattori et al., 1985).

More recent studies have compared GABA concentrations and GAD activity in parietal cortex and hippocampus of El mice that had been stimulated weekly for 25 weeks, El mice that had never received vestibular stimulation, and control ddY mice (Yoshiya et al., 1989). In parietal cortex, stimulated El mice had markedly lower GABA concentrations and GAD activity than ddY mice. Unstimulated El mice had marginally lower GAD activity but unaltered GABA levels compared to ddY mice. In the hippocampus, unstimulated El mice were similar to ddY mice, whereas stimulated El mice had higher GABA concentrations and increased GAD activity. This latter finding is compatible with the findings of an immunocytochemical study of the hippocampus of El mice that had experienced at least 10 seizures; the density of cells with GABA-like immunoreactivity was higher in stratum radiatum of El than C57BL/6 mice (King and La Motte, 1988).

BZ binding determined autoradiographically found no significant differences between unstimulated El and ddY mice, whereas in stimulated El mice BZ binding was significantly lower in parietal cortex, CA3 and CA4 regions of the hippocampus, the granular layer of the dentate gyrus and the amygdala (Shirasaka et al., 1989).

Other neurotransmitters in El mice. Acetylcholine concentration, choline acetyltransferase activity, and acetylcholine synthesis from glucose in vitro are higher in the brains of stimulated or unstimulated El mice than gpc or dd controls (Naruse et al., 1960; Kurokawa et al., 1966). Although there were no strain differences in cholinesterase activity, stimulated El mice had lower activity than non-stimulated El mice (Naruse et al., 1960). The density of muscarinic receptor binding sites did not differ between stimulated or unstimulated El, ddY, and C57BL/6 mice (Mori et al., 1983).

Seizure prone El mice have lower inter-ictal brain 5-HT concentrations, lower [3H]5-HT binding in the cortex and brain stem and lower K^+-evoked release of 5-HT from the cerebral cortex in vitro than non-stimulated El mice (Hiramatsu, 1981; 1983; Hiramatsu et al., 1986).

Administration of 5-hydroxtryptophan plus a monoamine oxidase inhibitor increased brain 5-HT concentration and reduced seizures (Hiramatsu, 1981). Inter-ictal concentrations of DA and NA are also lower in seizure-prone El mice than unstimulated El mice (Hiramatsu and Mori, 1977), as are the concentrations of glutamate and aspartate (Naruse et al., 1960; Kurokawa et al., 1966).

DBA/2 Mice

Background and phenomenology. In response to auditory stimulation DBA/2 mice develop characteristic seizures, consisting of an initial wild running phase terminated by the animal falling on one side, rhythmical clonic limb jerking progressing to tonic extension, which often results in respiratory arrest and death. Lesion studies indicate the importance of the central nucleus of the inferior colliculus, the deep superior colliculus and the adjacent tegmentum for the expression of audiogenic seizures in DBA/2 mice (Willott and Lu, 1980).

The audiogenic seizure susceptibility in DBA/2 mice is age-related, with maximal sensitivity between 21–28 days of age, with considerably reduced sensitivity before 21 days and almost complete resistance by 40 days of age. For neurochemical studies, DBA/2 mice at the age of maximum seizure susceptibility are compared with age-matched mice of a non-seizure susceptible strain. Some studies have also compared DBA/2 mice at ages before and after the age of maximum susceptibility with age-matched controls in an attempt to distinguish neurochemical differences that are related to seizure susceptibility from those that are purely strain related.

GABA in DBA/2 mice. A wide range of drugs that enhance GABA-mediated inhibition protect DBA/2 mice against audiogenic convulsions. These include direct acting $GABA_A$ agonists, such as muscimol (Anlezark et al., 1977), GABA-transaminase inhibitors, such as ethanolamine O-sulphate, γ-acetylenic GABA and γ-vinyl GABA (Anlezark et al., 1976; Schechter et al., 1977) and inhibitors of GABA reuptake mechanisms, particularly those selective for inhibiting glial reuptake (Horton et al., 1979; Croucher et al., 1983; Shousboe, this volume). These findings have stimulated the search for abnormalities in the GABA-neurotransmitter system of DBA/2 mice. Whole brain GABA concentration and GAD activity have not been found to differ between DBA/2 mice and other non-audiogenic seizure susceptible strains of mice (Tunnicliff et al., 1973; Sykes and Horton, 1982).

Radioligand binding to the $GABA_A$ receptor complex has been fairly extensively studied. GABA binding in whole brain of 3–4 month old DBA/2 mice was reported to be lower than C57BL/6 mice (Ticku, 1979). Horton et al., (1982) compared [^{3}H]GABA binding to whole brain homogenates of DBA/2 mice and TO mice (a strain much less seizure prone) at ages ranging from 8–43 days. The number of high affinity binding sites was lower, but the binding affinity higher, in DBA/2 mice at all ages. The number of low affinity [³H]GABA binding sites was lower in DBA/2 mice at 8–9 and 40–43 days of age, but did not differ during the period of maximum seizure susceptibility. Since apparent differences in high affinity [³H]GABA binding might arise because of the presence of endogenous compounds that can mask these sites, a further series of experiments were performed using a higher concentration of detergent and a more extensive washing procedure (Horton et al., 1984). Under these conditions, [³H]GABA binding was almost identical in cerebral cortex, forebrain, mid-brain, hippocampus; and cerebellum of DBA/2 and C57BL/6 mice at all ages studied. In pons-medulla, binding was essentially identical in DBA/2 and C57BL/6 mice at 13–15 days of age, when binding was described by a single component and at 40–43 days when there were clearly two components to the binding. However, at 21–23 days [³H]GABA binding in C57BL/6 mice was clearly best described by two components, while in DBA/2 mice only a single component was detected. In general agreement with these findings, Frandsen et al., (1987) reported [³H]GABA binding in frontal and occipital cortex and hippocampus not to differ in detergent treated membranes from 27 day old DBA/2 and C57BL/6 mice.

Robertson (1980) reported a high number of BZ binding sites in whole brain homogenates of DBA/2 mice at 21–24 days compared with C57BL/6 or BALB/C mice, whereas Kellogg and Syapin (1981) found no significant difference, and Horton et al. (1982) found a slightly lower number of BZ binding sites in DBA/2 compared to TO mice at 28–29 days but not at other ages. Using an autoradiographic method, Olsen et al. (1985) reported lower BZ binding in DBA/2 mice in several mid-brain and brain stem areas, including the substantia nigra and inferior colliculus. The ability of GABA to stimulate [³H]BZ binding did not differ between DBA/2 and TO mice up to 22–23 days of age, but was higher in DBA/2 mice at 28–29 and lower at 40–43 days (Horton et al., 1982).

The uptake of [³H]GABA into partially purified synaptosomes prepared from cerebral cortex, cerebellum, and pons-medulla of 13–43 day old DBA/2 and C57BL/6 mice revealed no differences in uptake velocity; Km was lower in the cerebellum of DBA/2 mice at 21–23 days of age and in the pons-medulla at 40–43 days of age (Spyrou et al., 1984). Cerebral cortical GABA uptake sites (labeled with [³H] nipecotic acid) also did not differ between between DBA/2 and C57BL/6 mice (Spyrou et al., 1984).

Other neurotransmitters in DBA/2 mice. Two studies have found NA concentration to be lower in whole brain (Schlesinger et al., 1965) and in forebrain and hindbrain (Kellogg, 1976) of DBA/2 mice than C57BL/6 mice at the time of maximal seizure susceptibility, but not at other times. However, Lints et al. (1980) found no difference in whole brain NA concentrations between DBA/2 and C57BL/6 at 16 and 28 days of age.

Graillot et al. (1985) and Jazrawi and Horton (1986) have examined noradrenergic receptor binding in brain homogenates of DBA/2 and

C57BL/6 mice at various ages. Whole brain α_1-adrenoceptor binding (using [^{3}H]prazosin) was significantly lower in DBA/2 mice by 13% at 24–26 days of age (Graillot et al., 1985) and by 17% at 21–23 days of age (Jazrawi and Horton, 1986), but did not differ at other ages. Graillot et al. (1985) reported lower α_1-adrenoceptor binding in DBA/2 mice was restricted to the olfactory bulbs (lower by 27%) and the cerebral cortex (lower by 17%). Jazrawi and Horton (1986) found lower α_1-adrenoceptor binding in several brain areas of DBA/2 mice, but in none was the differences statistically significant. Graillot et al. (1985) also reported significantly lower (by 15%) α_2-adrenoceptor binding (using [^{3}H]clonidine) in whole brain homogenates of DBA/2 than C57BL/6 mice at 24–26 days of age, but not at other ages. Binding in olfactory bulbs and cerebral cortex was lower by 53% and 23% respectively in DBA/2 mice, while in the brainstem binding was 17% higher. In contrast Jazrawi and Horton (1986) found no significant differences in whole brain [^{3}H]clonidine binding between DBA/2 and C57BL/6 mice between 8 and 43 days of age. At 21–23 days of age, [^{3}H]clonidine binding was 12–13% higher in the mid-brain and pons/medulla of DBA/2 mice; these differences were not statistically significant. β-Adrenoceptor binding sites (labeled with [^{3}H] dihydroalprenolol) have been reported not to differ in whole brain homogenates of DBA/2 and C57BL/6 mice between 7 and 43 days of age (Graillot et al., 1985; Jazrawi and Horton, 1986).

Schlessinger et al. (1965) reported brain 5-HT concentration to be lower in DBA/2 than C57BL/6 mice, at the time of peak susceptibility to audiogenic seizures, but not at 42 days of age when DBA/2 mice are no longer sensitive. Kellogg (1971) confirmed the lower brain concentration of 5-HT in DBA/2 mice at 21 days of age but also found that the lower concentration persisted at 42 days of age. At 21 days of age, 5-HT concentration and the rate of 5-HT synthesis were lower in forebrain of DBA/2 than C57BL/6 mice, whereas in hindbrain 5-HT concentration did not differ and 5-HT synthesis was higher in DBA/2 (Kellogg, 1976). In contrast to these findings, McGeer et al. (1969) and Lints et al. (1980) found no differences in brain 5-HT concentration between DBA/2 and C57BL/6 mice.

The number of 5HT$_2$ binding sites (labeled with [^{3}H]ketanserin) was 20% higher, with affinity unchanged, in the cerebral cortex of DBA/2 than in C57BL/6 mice at 21–23 days of age, but did not differ significantly at 13–15 or 40–45 days of age (Jazrawi and Horton, 1989). No significant differences were found in the number or affinity of 5HT$_2$ binding sites between DBA/2 and C57BL/6 mice in forebrain, mid-brain, hippocampus or pons-medulla at any ages (Jazrawi and Horton, 1989).

High affinity uptake of noradrenaline and serotonin into partially purified synaptosomes prepared from cerebral cortex, forebrain, mid-brain, hippocampus, and pons-medulla of DBA/2 and C57BL/6 mice, at ages prior to, during and following the period of maximum seizure susceptibility in DBA/2 mice, revealed no significant differences in affinity or maximum velocity (Jazrawi et al., 1987; Jazrawi and Horton, 1989).

Higher activity of choline acetyltransferase (frontal and temporal cortex) and acetylcholine esterase (temporal cortex) has been reported in DBA/2 than C57BL/6 mice (Ebel et al., 1973). The density of muscarinic acetylcholine receptors was significantly higher in the hippocampus of DBA/2 than C57BL/6 mice. This was most apparent at 14 days of age and much less marked at 42 days of age (Aronstam et al., 1979). Further, the proportion of receptors that exhibited a low affinity for muscarinic agonists was greater in DBA/2 mice at 14 and 21, but not at 42 days of age (Aronstam et al., 1979).

Although dopamine agonists prevent audiogenic seizures in DBA/2 mice (Anlezark and Meldrum, 1975; Anlezark et al., 1978), there is little direct neurochemical evidence to indicate an abnormality of dopamine function. Despite the substantial pharmacological evidence that antagonists at EAA receptors, particularly those of the N-methyl D-aspartate (NMDA) subclass, are able to block audiogenic seizures in DBA/2 mice (Croucher et al., 1982; Meldrum 1985; 1986), only quisqualic acid (QA) receptors have been compared in DBA/2 and C57BL/6 mice (Frandsen et al., 1987). Significantly higher QA binding (labeled with [^{3}H]AMPA) was reported in frontal cortex of DBA/2 mice at 21 and 27, but not 40, days of age. In the hippocampus, binding did not differ at 21 and 40 days of age but was significantly higher in DBA/2 than C57BL/6 mice at 27 days of age. In the cerebellum, binding was higher in DBA/2 mice at 27 and 40, but not 21 days of age. Although the significance of these findings to audiogenic seizure susceptibility in DBA/2 mice is not clear, the magnitude of these differences (more than 2-fold in the frontal cortex) is much greater than differences reported for other neurotransmitter receptors.

Epileptic Chickens

Background and phenomenology. Crawford (1970) described an autosomal recessive mutation in chickens resulting in high seizure susceptibility. Homozygotes (epileptic) exhibit spontaneous convulsions throughout their life span and seizures can be induced by intermittent photic stimulation (IPS) and hyperthermia. During IPS, the head and neck are slowly rotated and arched back and upward, followed by wing extension with some loss of balance, progressing to violent flapping of the wings and clonic movements of the legs. There is no tonic component to the seizure. The inter-ictal EEG is abnormal, characterized by slow wave, high voltage, spike and wave activity. During IPS, this is replaced by bilateral spiking synchronous with the stimulus. Following the seizure there is a period of post-ictal depression when the EEG is relatively silent and the bird is unresponsive to IPS (Johnson and Tuchek, 1987).

GABA in epileptic chickens.   GABA concentration has been reported to be higher in the cerebral hemispheres of adult epileptic than in heterozygous non-epileptic chickens, but not to differ in the optic lobes or cerebellum. In contrast GABA concentration did not differ between 1-day-old and 6-week-old epileptic and heterozygous chickens in any of the brain areas (Johnson and Tuchek, 1987). It is thus possible that the increased GABA concentration in adult chickens is an adaptive response to repeated seizures.

Adult epileptic chickens are remarkably more sensitive to the convulsant effects of β-carboline-3-carboxylate methyl ester (β-CCM) than their heterozygous littermates (Johnson et al., 1984). In the dose range 0.03–0.25 mg/kg, β-CCM induced convulsions in all epileptic chickens, whereas doses greater than 3mg/kg were required in their littermates, suggesting a possible abnormality in the GABA-receptor-ionophore complex. However, there was no difference in the ability of β-CCM to displace [^{3}H]diazepam binding in vitro from synaptosomal membranes prepared from the cerebral hemispheres of epileptic and heterozygous chickens. Binding of [^{3}H]diazepam or [^{3}H]flunitrazepam to membranes prepared from 2 day old or adult chickens, revealed no differences in the number or affinity of BZ binding sites or in the ability of GABA to enhance BZ binding between epileptic and heterozygous chickens (Fisher et al., 1985).

Other neurotransmitters in epileptic chickens.   Adult epileptic chickens have higher concentrations of NA and lower concentrations of 5-HT and DA in the cerebral hemispheres than non-epileptic chickens (Johnson et al., 1979), although the magnitude of the differences is relatively small. No differences in the concentrations of these monoamines were found in optic lobes, mid-brain/pons, medulla, and cerebellum. It seems unlikely that differences in monoamine concentrations contribute significantly to the high seizure susceptibility since administration of L-tryptophan, alone or in combination with a monoamine oxidase inhibitor, L-DOPA or adrenoceptor blocking drugs did not alter the IPS-induced susceptibility of chickens to seizures (Johnson et al., 1979).

There is some evidence to suggest abnormalities within the cholinergic system. The activity of acetylcholinesterase was found to be higher in cerebral hemispheres and cerebellum, and choline acetyl transferase activity lower in the cerebral hemispheres of epileptic chickens compared to their non-epileptic littermates (Johnson et al., 1979). It is not clear if these differences contribute to higher seizure susceptibility or are a consequence of repeated spontaneous seizures.

Spontaneous Petit Mal-Like Epilepsy in the Rat

Background and phenomenology.   Vergnes et al. (1982) reported spontaneous spike and wave discharges (SWD) in the cortical EEG of quiet, awake Wistar rats. The SWD had a mean frequency of 8 Hz, lasted for 1–90s (usually about 10s) and occurred at a mean rate of 1 per minute. SWD was associated with immobilisation of the rat with rhythmical clonus of the vibrissae and facial muscles. The SWD occurred over the entire neocortex but predominantly the fronto-parietal region and Vergnes et al. (1987) have provided evidence of their origin in lateral thalamic nuclei. Selective breeding has produced colonies in which SWD is present or absent in all animals. The SWD appear at 40–120 days of age and their duration increases with age (Vergnes et al., 1986). SWD are suppressed by benzodiazepines, valproate, trimethadione and ethosuximide and enhanced by phenytoin and carbamazepine (Micheletti et al., 1985). On the basis of these findings, spontaneous SWD in the rat has been proposed as a useful model of human petit mal (absence) epilepsy.

GABA in spontaneous petit mal epilepsy in the rat. The $GABA_A$ agonists, muscimol and THIP, markedly and dose-dependently increase the duration of the SWD, as did the $GABA_B$ agonist, baclofen, and the GABA-transaminase inhibitor, γ-vinyl GABA (Vergnes et al., 1984). Similar findings with muscimol have also been reported in the WAG/Rij rat model of petit mal epilepsy (Peeters et al., 1989). In both rat models, bicuculline reduced SWD. Direct GABA agonists have also been shown to induce or enhance SWD in other models of epilepsy (this chapter; Myslobodsky, 1984; Snead, 1984; Golden and Fariello, 1984). The contrast between the anticonvulsant effects of GABA agonists in generalised convulsive epilepsy (this chapter, and chapter 8) and their enhancement of SWD has prompted speculation that SWD is associated with GABA hyperactivity. However, other evidence suggests that the temporal and spacial actions of GABA are important. It may be of significance that GABA agonists injected bilaterally into the substantia nigra suppress SWD in the petit mal rat model (Depaulis et al., 1988), as well as inhibiting many other generalised or focally induced convulsions (Iadarola and Gale, 1982; Garant and Gale, 1983; McNamara et al., 1983; Millan et al., 1988).

Although there is little neurochemical data on the Wistar petit mal model, BZ binding has been studied. No differences in the binding of [^{3}H] flunitrazepam to the "central-type" BZ receptor were found in cortex, cerebellum , or hippocampus, compared with non-epileptic rats. However, the affinity of [^{3}H]Ro 5–4864 for the "peripheral-type" BZ binding site was markedly lower in cortex, cerebellum and hippocampus of petit mal rats, with no differences in the number of binding sites (Hariton et al., 1988). The functional significance of these sites remains unclear, although Ro 5–4864 has been shown to facilitate audiogenic seizures in DBA/2 mice and, at higher doses, to induce seizures in the absence of auditory stimulation (Bénavidès et al., 1984).

Photosensitive Epilepsy in the Baboon

Baboons from the Casamance region of Senegal have a high naturally-occurring susceptibility to photically-induced epilepsy. Intermittent photic stimulation (IPS) produces myoclonus, initially of the eyelids and progressing sequentially to the face, head, trunk, and limbs. The myoclonus may become self sustaining after the discontinuation of stimulation and lead to tonic-clonic seizures. Myoclonus of the eyelids is associated with EEG abnormalities primarily in the fronto-rolandic cortex; limb myoclonus is synchronous with diffuse cortical spike and spike and wave activity.

Although there are few clues to the neurochemical basis of this primate model of epilepsy, myoclonic response are modified by pharmacological manipulation of several neurotransmitter systems. Some drugs which enhance GABA-mediated transmission are effective in this model: benzodiazepines are very potent and GABA-transaminase inhibitors produce long lasting protection against ILS-induced seizures (Meldrum and Horton, 1978; Meldrum and Braestrup, 1984). In contrast, the potent direct acting GABA agonists, muscimol and THIP, do not inhibit myoclonic responses but produce paroxysmal EEG abnormalities, sometimes associated with diffuse myoclonus of the limbs (Pedley et al., 1979; Meldrum and Horton, 1980). Similar effects have been reported in other models of epilepsy (see spontaneous petit mal in the rat, this chapter). GABA, glutamate and aspartate concentrations were found not to differ in the frontal lobes of baboons that had shown consistent epileptic abnormalities for 3 years compared to non-epileptic baboons (Hansen et. al., 1973). Cortical BZ, muscarinic acetylcholine and opiate binding did not differ in a spontaneously epileptic baboon compared to moderately photosensitive or non-photosensitive baboons. BZ binding in the spontaneously epileptic baboon was, however, more thermolabile in vitro than in the other baboons (Squires et al., 1979). Other drugs which block photosensitive epilepsy in the baboon include excitatory amino acid antagonists, particularly of the NMDA subtype (Meldrum, 1985; 1986), DA agonists (Anlezark et al., 1981), 5-HT agonists and 5-hydroxytryptophan (Wada et al., 1972).

INDUCTIVE MODELS OF EPILEPSY
Cobalt-Induced Epilepsy in the Rat

Background and phenomenology. Unilateral application of metallic cobalt to the rat motor cortex produces a semi-chronic model of focal epilepsy with distinct evolutionary phases. Following implantation the EEG remains normal for a few days, followed by localized EEG spiking associated with myoclonic limb jerking. Gradually the motor signs abate, although the EEG spiking activity continues and finally the EEG spiking disappears. The period of peak seizure activity lasts 7–10 days and few seizures are seen later than the second week after implantation. The focal nature of the seizures and the predictable time course make this a suitable model to study biochemical events that precede seizure onset and that are the results of repeated seizures.

GABA and cobalt-induced epilepsy in the rat. GABA concentration and GAD activity were markedly reduced (to 47% and 25% of control values, respectively) 7 days after cobalt implantation, at the time of peak seizure activity (Ross and Craig, 1981a). Both had returned to control values by 21 days, when the seizures had stopped. The maximum velocity of high affinity GABA uptake is also decreased, with affinity unchanged, in the epileptic focus (Balcar et al., 1978; Ross and Craig, 1981b). The maximal decrease (to 34% of controls) occurred 7 days after cobalt implantation. A gradual return to control values was seen from 14 days onwards, although 35 days after cobalt implantation uptake remained 15% below control values (Ross and Craig, 1981b). Eight to ten days after cobalt implantation, the Vmax of GABA uptake in the focus was reduced to 27% of control values, whereas the Km and Vmax of high-affinity L-glutamate uptake was unaltered (Balcar et al., 1978). The number of [^{3}H]GABA binding sites was increased markedly at 7 and 14 days after implantation; at 21 days they had returned towards control values (Ross and Craig, 1981a). Although it is not clear from the above study, Esclapez and Trottier (1989) using the same model have demonstrated a reduction in GABA-immunoreactive cells and terminals in the perilesional area before the onset of epileptic discharges. The loss of GABA-immunoreactivity became more marked during the period of maximal spiking activity, and progressively returned to normal values as the seizures waned, except in the deep cortical layers closest to the lesion. Thus both biochemical and morphological studies indicate that reductions in presynaptic GABA markers precede seizure activity, exhibit maximal reductions that relate to the severity of seizures, and are largely reversed following the termination of seizures.

Other neurotransmitters in cobalt-induced epilepsy in the rat. Decreased concentrations of NA and NA metabolites and decreased NA uptake also occur in the perifocal area, with the greatest decreases occurring 8–14 days after implantations and a return to normal by 40 days, when seizure activity had ceased (Trottier et al., 1981; 1983). Effects on the serotonergic system were less pronounced and of shorter duration than those on the noradrenergic system. 5-HT concentration and 5-HT uptake were lower 8–10 days after cobalt implantation, with a tendency for 5-HIAA concentrations to be increased. At 20 days, 5-HT and 5-HIAA concentrations and 5-HT uptake had returned to control levels; NA concentration and NA uptake remained significantly reduced at this time (Trottier et al., 1983).

Decreased activity of choline acetyltransferase and acetylcholinesterase and decreased concentrations of acetylcholine have also been reported in the perifocal area; again maximal reductions were seen at 7–14 days and had returned to control levels by 21 days (Goldberg et al., 1972; Emson and Joseph, 1975; Hoover et al., 1977).

Aluminium-Induced Epilepsy in the Monkey

Background and phenomenology. Implantation of metallic aluminum or injection of aluminum hydroxide gel unilaterally into the sensorimotor cortex of monkeys produces a stable chronic model of intermittent focal epilepsy. A focus of spike and wave discharges develops in the area of cortex surrounding the aluminium granuloma. The time of onset of seizures is 2–3 months; when fully established they continue to occur spontaneously throughout the life of the animal.

GABA and aluminium-induced epilepsy in the monkey. Bakay and Harris (1981) performed a neurochemical analysis of tissue from monkeys who had aluminium-induced focal seizures over a two year period. GABA concentration and GAD activity were markedly decreased in the aluminium granuloma and the epileptic focus, with normal values observed as the distance from the focus increased. Within the granuloma there was a marked and rather uniform loss of neurotransmitter receptor binding sites (including GABA$_A$, α- and β-adrenoceptors, glutamate and muscarinic acetylcholine), consistent with the neuronal cells loss within the granuloma. In the epileptic focus, the loss of GABA$_A$ receptor binding was more marked (60%) than that of other neurotransmitter receptor binding sites (about 30%), suggesting some selective loss of cells with GABA$_A$ receptors.

A selective loss of GAD-positive nerve terminals and cell bodies has been demonstrated in monkeys sacrificed at relatively short intervals following aluminium implantation, before the development of clinical seizures (Houser et al., 1986; Ribak et al., 1986; 1989). Loss of GAD-positive terminals was progressive with time following the aluminium implants and the development of EEG and clinical seizures. In the focus, there was a marked loss of GAD-positive terminals that synapse on pyramidal neurons. Terminals from two cell types have been clearly identified; the basket cells which form axosomatic synapses and the chandelier cells which form symmetrical synapses with the axon initial segments of the pyramidal neurons (Ribak et al., 1982; Ribak, 1985). The loss of these GABA-mediated inhibitory inputs is likely to contribute to the increased excitability of pyramidal neurons.

Kindling

Background and phenomenology. Kindling is a phenomenon in which repeated (usually daily) administration of focal low intensity, high frequency sub-convulsive electrical stimulation results in a progressive intensification of the seizure, eventually resulting in generalized motor convulsions that may subsequently occur spontaneously. When motor seizures develop the animal is said to be kindled. Kindling produces a long-term change in sensitivity; kindled animals unstimulated for as long as 12 months develop motor seizures with one or two subsequent stimulations. Kindling can be induced in many brain areas, particularly those associated with the limbic system (amygdala, hippocampus, entorhinal, and pyriform cortex). Kindling can be induced in a wide range of species; the rat has been most widely studied.

Pharmacological and neurochemical studies have focused on different aspects of kindling; those relating to the induction of kindling (the number of stimuli required for kindling), to the effects of repeated seizures (studies performed within a few days of the establishment of kindling or after a fixed number of kindled seizures) and, of most interest, to those responsible for the long-term change in seizure susceptibility (effects that persist several weeks after the last seizure).

GABA and kindling. Drugs which enhance GABA-mediated inhibition (direct acting GABA agonists, benzodiazepines, barbiturates and GABA-transaminase inhibitors) inhibit the establishment of kindling and suppress fully kindled seizures (McNamara et al., 1987). The development of kindling in the rat hippocampus has been shown to be associated with an increase in the number of GABA-immunoreactive somata in the CA1 region (Kamphuis et al., 1987). The cell density on the stimulated side was higher by 10% compared to controls following 6 afterdischarges, and by 38%, with enhanced labeling density, after 14 afterdischarges. However, in fully kindled rats (34 afterdischarges) there were no significant differences compared to controls.

Shin et al. (1985) studied GABA and BZ binding in rats 24h and 28 days following 3 amygdala and entorhinal cortex kindled seizures. Twenty-four hours after the last seizure, GABA binding was increased by 22% and BZ binding by 38% in the dentate gyrus compared to controls, but did not differ in neocortex or substantia nigra. The increased binding was localized to the soma and dendritic fields of the granule cells. Twenty-eight days after the last seizure no differences in GABA or BZ binding were found in the dentate gyrus or the other brain areas. The authors suggested that the transient nature of the changes was an adaptive response to repeated seizures; i.e., an attempt to stabilize granule cell excitability and reduce the likelihood of further seizures. Compatible with this is the finding of enhanced K^+-stimulated GABA release from the hippocampus 24h after 5–7 generalized seizures kindled from entorhinal cortex (Liebowitz et al., 1978). No differences in GABA concentration or GAD activity were found in 5 brain regions (including hippocampus and amygdala) 1 week after 19–24 amygdala kindled seizures (Lerner-Natoli et al., 1985).

Studies performed at longer time intervals after a series of kindled seizures have produced equivocal findings. Two weeks following 6 amygdala kindled seizures, [^{3}H]GABA and [^{3}H]muscimol binding did not differ significantly in 12 brain regions of kindled and control rats (Tuff et al., 1983). However, the number of BZ binding sites was significantly higher in hippocampus and amygdala and significantly lower in the caudate of kindled rats. GABA stimulation of BZ binding was unaltered. Niznik et al. (1983) also studied BZ binding 2 weeks after the last of 6 amygdala kindled seizures. In contrast to Tuff et al. (1983), no differences in BZ binding were found in amygdala, hippocampus, or striatum but BZ binding was lower (by 20%) in the frontal cortex and hypothalamus. A possible explanation for these discrepant findings may relate to endogenous modulators of BZ binding sites. Niznik et al. (1983) used crude homogenates, whereas Tuff et al. (1983) studied extensively washed membrane preparations. Nishimura et al. (1988) studied rats 4 weeks after a series of amygdala kindled seizures. GABA concentrations and

GABA$_A$, GABA$_B$ and BZ binding did not differ in 6 brain regions (including amygdala and hippocampus) of kindled and control rats.

The lack of consistent findings at long intervals after kindled seizures may relate to the relatively large brain areas studied. Kamphius et al., (1989) have provided immunocytochemical evidence for a long-term decrease in GABA concentration in the CA1 region of the hippocampus 4 weeks after the last of 10 hippocampal kindled seizures.

Other neurotransmitters in kindling. Depletion of brain NA facilitates kindling (McNamara et al., 1987). However, NA mechanisms do not appear to be critically involved in the maintenance of enhanced sensitivity. Although brain NA concentrations are decreased at short intervals following kindling (Callaghan and Schwark, 1979), NA concentrations and K$^+$-evoked release of NA are unaltered at longer time intervals (Kant et al., 1980; Blackwood, 1981). β-Adrenoceptors have been shown to decrease transiently following kindled seizures (McNamara, 1978), although McIntyre and Roberts (1983) reported decreased ligand binding 23 days after the last kindled seizure. Available evidence suggests that serotonin and dopamine mechanisms are not essential for the development or permanence of kindling (McNamara et al., 1987).

Although pharmacological evidence suggests cholinergic mechanisms are not essential for the development or expression of kindled seizures, cholinergic markers are altered by kindling. Noda et al. (1982) reported decreased activity of choline acetyltransferase in frontal cortex, hippocampus and amygdala 2 weeks after the final amygdala-kindled seizure. Muscarinic receptors are also decreased in the brains of kindled rats, although this effect is not sustained. (Noda et al., 1982; Blackwood et al., 1982). McNamara et al. (1987) suggest that the alterations in cholinergic markers are a result of repeated seizures rather than being related to the permanence of kindling.

Antagonists of the NMDA-type EAA receptors are able to suppress kindled seizures (Peterson et al., 1983) and there is evidence for increased quisqualate-type binding (Savage et al., 1984b) and decreased kainate-type (Savage et al., 1984a) EAA binding following kindled seizures, although at least for the quisqualate-type receptor the effect is transient. Long-term increases in dynorphin A concentration (Przewlocki et al., 1983) localized to the hippocampus, and more widespread increases in somatostatin immunoreactivity (Kato et al., 1983) are possible, but as yet unproven, biological substrates for the permanence of kindling.

CONCLUSIONS

For almost all animal models of epilepsy there is direct or indirect evidence for abnormalities of neurochemical markers of GABA-mediated neurotransmission. In some models (focal epilepsy induced by cobalt in the rat or aluminium in the monkey), the evidence indicates that loss of presynaptic GABA markers precedes, and thus probably contributes substantially to the epileptic discharges. This is conceptually easy to rationalise with a loss of GABA-mediated inhibitory control resulting in excessive burst firing. In other models (epileptic chickens, epileptic gerbils, and GEPRs) presynaptic GABA markers are increased. In the case of epileptic chickens and gerbils increased GABA concentrations and increased GAD-positive neurons may be adaptive responses to repeated spontaneous seizures (possibly unobserved or without motor signs), or to deficits in post-synaptic GABA receptor mechanisms. In the GEPR increased GAD-positive neurons do not appear to be related to previous seizure history. There is evidence suggesting reduced efficacy of GABA receptors in the GEPR (Faingold, 1988). An alternative explanation suggested by Roberts et al. (1985a, b) is an abnormal neuronal circuitry such that GABA neurons inhibit other GABAergic neurons thus causing disinhibition.

It is also clear that in the majority of animal models abnormalities in GABA markers are not the only neurochemical abnormalities related to transmitter function. There is considerable evidence for a major influence of noradrenaline and increasing evidence for abnormal EAA-mediated neurotransmission in many animal models of epilepsy.

The scope for further studies in existing models and the development of new models remains enormous. Immunocytochemical and autoradiographic techniques are important approaches that embrace the detailed anatomical localization of neurotransmitters and their receptors that are likely to prove increasingly important in the future.

REFERENCES

Anlezark GM, Meldrum BS: Effects of apomorphine, ergocornine and pirebedil on audiogenic seizures in DBA/2 mice. Br J Pharmacol 53: 419–426, 1975.

Anlezark G, Horton RW, Meldrum BS, Sawaya MCB: Anticonvulsant action of ethanolamine O-sulphate and di-n-propylacetate and the metabolism of γ-aminobutyric acid (GABA) in mice with audiogenic seizures. Biochem Pharmacol 25: 413–471, 1976.

Anlezark G, Marrosu F, Meldrum B: Dopamine agonists in reflex epilepsy. In: Morselli PL, Lloyd KG, Löscher W, Meldrum B, Reynolds EH (eds) "Neurotransmitters, Seizures and Epilepsy" New York: Raven Press, 1981, pp 251–262.

Anlezark G, Collins J, Meldrum BS: GABA agonists and audiogenic seizures. Neurosci Lett 7: 337–340, 1977.

Anlezark GM, Horton RW, Meldrum BS: Dopamine agonists and audiogenic seizures: The relationship between protection against seizures and changes in monoamine metabolism. Biochem Pharmacol 27: 2821–2828, 1978.

Aronstam RS, Kellogg C, Abood LG: Development of muscarinic cholinergic receptors in in-bred strains of mice: Identification of receptor heterogeneity and relation to audiogenic seizure susceptibility. Brain Res 162: 231–241, 1979.

Bakay RAE, Harris AB: Neurotransmitter, receptor and biochemical changes in monkey cortical epileptic foci. Brain Res 206: 387–404, 1981.

Balcar VJ, Pumain R, Mark J, Borg J, Mandel P: GABA-mediated inhibition in the epileptogenic focus, a process which may be involved in the mechanism of the cobalt-induced epilepsy. Brain Res 154: 182–185, 1978.

Bénavidès J, Guilloux F, Allam DE, Uzan A, Mizoule J, Renault C, Dubroeucq MC, Guérémy C, Le Fur G: Opposite effects of an agonist, Ro 5–4864, and an antagonist, PK 11195, of the peripheral type benzodiazepine binding sites on audiogenic seizures in DBA/2J mice. Life Sci 34: 2613–2620, 1984.

Blackwood D: The role of noradrenaline and dopamine in amygdaloid kindling. In Morselli PL, Lloyd KG, Löscher W, Meldrum B, Reynolds EH (eds): "Neurotransmitters, Seizures and Epilepsy." New York: Raven Press, 1981, pp. 203–211.

Blackwood DHR, Martin MJ, Howe JG: A study of the role of the cholinergic system in amygdaloid kindling in rats. Psychopharmacology 76: 66–69, 1982.

Booker JG, Dailey JW, Jobe PC, Lane JD: Cerebral cortical GABA and benzodiazepine binding sites in genetically seizure prone rats. Life Sci 39: 799–806, 1986.

Browning RA, Rigler-Daugherty SK, Long G, Jobe PC: Decreased norepinephrine (NE) uptake in cerebral cortex and inferior colliculus of genetically epilepsy prone (GEP) rats. Fed Proc 45: 675, 1986.

Callaghan DA, Schwark WS: Involvement of catecholamines in kindled amygdaloid convulsions in the rat. Neuropharmacology 18: 541–545, 1979.

Chapman AG, Faingold CL, Hart GP, Bowker HM, Meldrum BS: Brain regional amino acid levels in seizure susceptible rats: Changes related to sound-induced seizures. Neurochem Int 8: 273–279, 1986.

Crawford RD: Epileptiform seizures in domestic fowl. J Hered 61: 185–188, 1970.

Croucher MJ, Collins JF, Meldrum BS: Anticonvulsant action of excitatory amino acid antagonists. Science 216: 899–901, 1982.

Croucher MJ, Meldrum BS, Krogsgaard-Larsen, P: Anticonvulsant activity of GABA uptake inhibitors and their prodrugs following central or systemic administration. Eur J Pharmacol 89: 217–228, 1983.

Dailey JW, Jobe PC: Indices of noradrenergic function in the central nervous system of seizure-naive genetically epilepsy-prone rats. Epilepsia 27: 665–670, 1986.

Dailey JW, Reigel CE, Mishra PK, Jobe PC: Neurobiology of seizure predisposition in the genetically epilepsy-prone rat. Epilepsy Res 3: 3–17, 1989.

Depaulis A, Vergnes M, Marescaux C, Lannes B, Warter J-M: Evidence that activation of GABA receptors in the substantia nigra suppresses spontaneous spike-and-wave discharge in the rat. Brain Res 448: 20–29, 1988.

Ebel A, Hermetet JC, Mandel P: Comparative study of acetylcholinesterase and choline acetyltransferase enzyme activity in brain of DBA and C57 mice. Nature 242: 56–58, 1973.

Emson PC, Joseph MH: Neurochemical and morphological changes during the development of cobalt-induced epilepsy in the rat. Brain Res 93: 91–110, 1975.

Esclapez M, Trottier S: Changes in GABA-immunoreactive cell density during motor focal epilepsy induced by cobalt in the rat. Exp Brain Res 76: 369–385, 1989.

Faingold CL: The genetically epilepsy-prone rat. Gen Pharmacol 19: 331–338, 1988.

Faingold CL, Gehlbach G, Caspary DM: Decreased effectiveness of GABA-mediated inhibition in the inferior colliculus of the genetically epilepsy-prone rat. Exp Neurol 93: 145–159, 1986.

Faingold CL, Millan MH, Boersma CA, Meldrum BS: Excitant amino acids and audiogenic seizures in the genetically epilepsy-prone rat. 1. Afferent seizure initiation pathway. Exp Neurol 99: 678–686, 1988.

Fisher TE, Davis LG, Tuchek JM, Johnson DD, Crawford RD: Benzodiazepine receptors and seizure susceptibility in epileptic fowl. Can J Physiol Pharmacol 63:85–88, 1985.

Frandsen A, Belhage B, Shousboe A: Differences between seizure-prone and non-seizure-prone mice with regard to glutamate and GABA receptor binding in the hippocampus and other regions of the brain. Epilepsy Res 1:107–113, 1987.

Garant, DS, Gale K: Lesions of substantia nigra protect against experimentally induced seizures. Brain Res 273: 156–161, 1983.

Goldberg AM, Pollock JJ, Hartman ER, Craig CR: Alterations in cholinergic enzymes during the development of cobalt-induced epilepsy in the rat. Neuropharmacology 11: 253–259, 1972.

Golden GT, Fariello RG: Epileptogenic action of some direct GABA agonists: Effects of manipulation of the GABA and glutamate systems. In Fariello RG, Morselli PL, Lloyd KG, Quesney LF, Engel J (eds): "Neurotransmitters, Seizures and Epilepsy II." New York: Raven Press, 1984, pp 237–244.

Graillot C, Baumann N, Maurin Y: Modulation of α_1- and α_2- adrenoceptor binding sites in the brain of audiogenic seizure susceptible mice (DBA/2J). Eur J Pharmacol 118: 231–237, 1985.

Hansen S, Perry TL, Wada JA, Sokol M: Brain amino acids in baboons with light-induced epilepsy. Brain Res 50: 480–483, 1973.

Hariton C, Ciesielski L, Cano J-P, Mandel P: Studies on benzodiazepine receptor subtypes in a model of chronic spontaneous petit mal-like seizures. Neurosci Lett 84: 97–102, 1988.

Hattori H, Ito M, Mikawa H: γ-Aminobutyric acid, benzodiazepine binding sites and γ-aminobutyric acid concentrations in epileptic El mouse brain. Eur J Pharmacol 119: 217–223, 1985.

Hiramatsu, M: Brain monoamine levels and El mouse convulsions. Folia Psychol Neurol Jpn 35: 261–266, 1981.

Hiramatsu M: Brain 5-hydroxytryptamine level, metabolism and binding in El mice. Neurochem Res 8: 1163–1175, 1983.

Hiramatsu M, Kabuto H, Mori A: [^{3}H] 5-Hydroxytryptamine release from brain slices of El mice. Neurochem Res 11: 103, 1986.

Hiramatsu M, Mori A: Brain catecholamine concentrations and convulsions in El mice. Folia Psychol Neurol Jpn 31: 491–495, 1977.

Hoover DB, Craig CR, Colasanti BK: Cholinergic involvement in cobalt-induced epilepsy in the rat. Exp Brain Res 29: 501–513, 1977.

Horton RW, Collins JF, Anlezark GM, Meldrum BS: Convulsant and anticonvulsant actions in DBA/2 mice of compounds blocking the reuptake of GABA. Eur J Pharmacol 59: 75–83, 1979.

Horton RW, Prestwich SA, Meldrum BS: γ-Aminobutyric acid and benzodiazepine binding sites in audiogenic seizure-susceptible mice. J Neurochem 39: 864–870, 1982.

Horton RW, Prestwich SA, Meldrum BS: Neurotransmitter receptor binding in genetically seizure-susceptible mice. In Fariello RG, Morselli PL, Lloyd KG, Quesney LF, Engel J (eds): "Neurotransmitters, Seizures and Epilepsy II." New York: Raven Press, 1984, pp 191–200.

Houser CR, Harris AB, Vaughn JE: Time course of the reduction of GABA terminals in a model of focal epilepsy: A glutamic acid decarboxylase immunocytochemical study. Brain Res 383: 129–145, 1986.

Huxtable RJ, Laird H, Bonhaus D, Thies AC: Correlations between amino acid concentrations in brains of seizure-susceptible and seizures-resistant rats. Neurochem Int 4: 73–78, 1982.

Iadarola M, Gale K: Substantia nigra: site of anticonvulsant activity mediated by γ-aminobutyric acid. Science 218: 1237–1240, 1982.

Jazrawi SP, Horton RW: Brain adrenoceptor binding sites in mice susceptible (DBA/2J) and resistant (C57BL/6) to audiogenic seizures. J Neurochem 47: 173–177, 1986.

Jazrawi SP, Horton RW: 5-HT$_2$ receptor binding and 5-HT uptake in mouse brain: developmental changes and the relationship to audiogenic seizure susceptibility in DBA/2J mice. Dev Brain Res 45:257–263, 1989.

Jazrawi SP, Yamaguchi Y, Horton RW: Synaptosomal high-affinity noradrenaline uptake does not differ between mice susceptible (DBA/2J) and resistant (C57BL/6) to audiogenic seizures. Epilepsy Res 1: 139–141, 1987.

Jobe PC: Pharmacology of audiogenic seizures. In Brown RD, Daigneault EA (eds): "Pharmacology of Hearing: Experimental and Clinical Bases." New York: Wiley, 1981, pp 271–304.

Johnson DD, Davis HL, Crawford RD: Pharmacological and biochemical studies in epileptic fowl. Fed Proc 38: 2417–2423, 1979.

Johnson DD, Fisher TE, Tuchek JM, Crawford Rd,: Pharmacology of methyl-and propyl-β-carbolines in a hereditary model of epilepsy. Neuropharmacololgy 23: 1015–1017, 1984.

Johnson DD, Tuchek JM: The epileptic chickens. In Jobe PC, Laird HE (eds): "Neurotransmitters and Epilepsy." Clifton, New Jersey: Humana Press, 1987, pp 95–114.

Kamphuis W, Wadman WJ, Buijs RM, Lopes Da Silva FH: The development of changes in hippocampal GABA immunoreactivity in the rat kindled model of epilepsy: A light microscopic study with GABA antibodies. Neuroscience 23:433–446, 1987.

Kamphuis W, Huisman E, Wadman WJ, Lopes Da Silva FH: Decrease in GABA immunoreactivity and altered GABA metabolism after kindling in the rat hippocampus. Exp Brain Res 74: 375–386, 1989.

Kant GJ, Meyerhoff JL, Corcoran ME: Release of norepinephrine and dopamine from brain regions of amygdaloid kindled rats. Exp Neurol 70: 701–705, 1980.

Kato N, Higuchi T, Friesen HG, Wada JA: Changes of immunoreactive somatostatin and β-endorphin content in rat brain after amygdaloid kindling. Life Sci 32: 2415–2422, 1983.

Kellogg C: Serotonin metabolism in the brains of mice sensitive or resistant to audiogenic seizures. J Neurobiol 2: 209–219, 1971.

Kellogg C: Audiogenic seizures: Relation to age and mechanisms of monoamine neurotransmission. Brain Res 106: 87–103, 1976.

Kellogg C, Syapin P: Development of benzodiazepine and GABA binding in inbred mice in relation to audiogenic seizures. Am Soc Neurosci 7: 847, 1981.

King JT, La Motte CC: VIP-, SS-, and GABA-like immunoreactivity in mid hippocampal region of El (epileptic) and C57BL/6 mice. Brain Res 475: 192–197, 1988.

Ko HK, Dailey JW, Jobe PC: Evaluation of monoaminergic receptors in the genetically epilepsy prone rat. Experientia 40: 70–73, 1984.

Kurokawa M, Naruse H, Kato M (1966): Metabolic studies on ep mouse, a special stain with convulsive predisposition. Prog Brain Res 21A: 112–130, 1966.

Laird HE, Jobe PC: The genetically epilepsy-prone rat. In Jobe PC, Laird HE (eds): "Neurotransmitters and Epilepsy." Clifton, New Jersey: Humana Press, 1987, pp 57–94.

Lee RJ, Bajorek JG, Lomax P: Similar anticonvulsant, but unique, behavioural effects of opioid agonists in the seizure-sensitive Mongolian gerbil. Neuropharmacology 23: 517–524, 1984.

Lee RJ, McCabe RT, Wamsley JK, Olsen RW, Lomax P: Opioid receptor alterations in a genetic model of generalized epilepsy. Brain Res 380: 76–82, 1986.

Lehmann A, Sandberg M, Huxtable R: In vivo release of neuroactive amines and amino acids from the hippocampus of seizure-resistant and seizure-susceptible rats. Neurochem Int 8: 513–520, 1986.

Lerner-Natoli M, Heaulme M, Leyris R, Biziere K, Rondouin G: Absence of modifications in γ-aminobutyric acid metabolism after repeated generalised seizures in amygdala-kindled rats. Neurosci Lett 62:271–276, 1985.

Liebowitz NR, Pedley TA, Cutler RWP: Release of γ-aminobutyric acid from hippocampal slices of the rat following generalised seizures induced by daily electrical stimulation of entorhinal cortex. Brain Res 138: 369–373, 1978.

Lints CE, Willott JF, Sze PY, Nenja LH: Inverse relationship between whole brain monoamine levels and audiogenic seizure susceptibility in mice: Failure to replicate. Pharmacol Biochem Behav 12: 385–388, 1980.

Lomax P, Lee RJ, Olsen RW: The spontaneously epileptic Mongolian gerbil. In Jobe PC, Laird HE (eds): "Neurotransmitters and Epilepsy." Clifton, New Jersey: Humana Press, 1987, pp 41–56.

Löscher W: Genetic animal models of epilepsy as a unique resource for the evaluation of anti-convulsant drugs. A review. Methods Find Exp Clin Pharmacol 6: 531–547, 1984.

Löscher W: Influence of pharmacological manipulation of inhibitory and excitatory neurotransmitter systems on seizure behaviour in the Mongolian gerbil. J Pharmacol Exp Ther 233: 204–213, 1985.

Löscher W, Frey HH, Reiche R, Schultz D: High anticonvulsant potency of γ-aminobutyric acid (GABA) mimetic drugs in gerbils with genetically determined epilepsy. J Pharmacol Exp Ther 226: 839–844, 1983.

McGeer EG, Ikeda H, Asakura T, Wada JA: Lack of abnormality in brain aromatic amines in rats and mice susceptible to audiogenic seizure. J Neurochem 16: 945–950, 1969.

McIntrye DC, Roberts DCS: Long-term reduction in β-adrenergic receptor binding after amygdala kindling in rats. Exp Neurol 82: 17–24, 1983.

McNamara JO: Selective alterations of regional β-adrenergic receptor binding in the kindling model of epilepsy. Exp Neurol 61: 582–591; 1978.

McNamara JO, Bonhaus DW, Crain BJ, Gellman RL, Shin C: Biochemical and pharmacological studies of neurotransmitters in the kindling model. In Jobe PC, Laird HE (eds): "Neurotransmitters and Epilepsy." Clifton, New Jersey: Humana Press, 1987, pp 115–160.

McNamara JO, Rigsbee LC, Galloway MT: Evidence that substantia nigra is crucial to neural network of kindled seizures. Eur J Pharmacol 86: 485–486, 1983.

Meldrum BS: Possible therapeutic applications of antagonists of excitatory amino acid neurotransmitters. Clin Sci 68: 113–122, 1985.

Meldrum B: Is epilepsy a disorder of excitatory transmission? In Trimble MR, Reynolds EH (eds): "What is Epilepsy" Edinburgh: Churchill Livingston, 1986, pp 293–302.

Meldrum B, Braestrup C: GABA and the anticonvulsant action of benzodiazepines and related drugs. In Bowery NG (ed): "Actions and Interactions of GABA and Benzodiazepines." New York: Raven Press, 1984, pp 133–153, 1984.

Meldrum BS, Horton RW: Blockade of epileptic responses in the photosensitive baboon, Papio papio, by two irreversible inhibitors of GABA-transaminase γ-acetylenic GABA (4-amino-hex-5-ynoic acid) and γ-vinyl GABA (4-amino-hex-5-enoic acid). Psychopharmacology 59: 47–50, 1978.

Meldrum BS, Horton RW: Effects of the bicyclic GABA agonist, THIP, on myoclonic seizure responses in mice and baboons with reflex epilepsy. Eur J Pharmacol 61: 231–237, 1980.

Micheletti G, Vergnes M, Marescaux C, Reis J, Depaulis A, Rumbach L, Warter JM: Antiepileptic drug evaluation in a new animal model: spontaneous petit mal epilepsy in the rat. Arzneim-Forsch Drug Res 35:483–485, 1985.

Millan MH, Meldrum BS, Boersma CA, Faingold CL: Excitant amino acids and audiogenic seizures in the genetically epilepsy-prone rat. II. Efferent seizure propagating pathway. Exp Neurol 99: 687–698, 1988.

Mills SA, Reigel CE, Jobe PC, Savage DD: Increase in the number of hippocampal [³H] glutamate binding sites in rats inbred for a genetic predisposition to acoustic stimulus induced seizures. Am Soc Neurosci 11: 1315; 1985.

Mimaki T, Yabuuchi H, Laird H, Yamamura HI: Effects of seizures and antiepileptic drugs on benzodiazepine receptors in rat brain. Ped Pharmacol 4: 205–211, 1984.

Mori A, Fujita S, Kwawguchi K, Chikami T, Ohno K, Tsuji M, Onishu H, Yamagami S, Kawakita Y: Muscrinic cholinergic receptors of the convulsive strain (El) mouse. Folia Psychiat Neurol Jpn 37: 299–302, 1983.

Myslobodsky MS: Petit mal epilepsy: To little or too much GABA? In Fariello RG, Morselli PL, Lloyd KG, Quesney LF, Engel J (eds): "Neurotransmitters, Seizures and Epilepsy II." New York: Raven Press, 1984, pp 337–346, 1984.

Naruse H, Kato M, Kurokawa M, Haba R, Yabe T: Metabolic defects in a convulsive strain of mouse. J Neurochem 5: 359–369, 1960.

Nicoletti F, Barbaccia ML, Iadarola MJ, Pozzi O, Laird HE: Abnormality of α₁-adrenergic receptors in the frontal cortex of epileptic rats. J Neurochem 46: 270–273, 1986.

Nishimura S, Yagi K, Ishida S, Seino M, Miyamoto K: No alteration of monoamines and GABA systems in the kindled rat brain. Japn J Psychiatr Neurol 42: 628–639, 1988.

Niznik HB, Kish SJ, Burnham WM: Decreased benzodiazepine receptor binding in amygdala-kindled rat brains. Life Sci 33: 425–430, 1983.

Noda Y, Wada JA, McGeer EG: Lasting influence of amygdaloid kindling on cholinergic neurotransmission. Exp Neurol 78: 91–98, 1982.

Olsen RW, Jones I, Seyfried TN, McCabe RT, Wamsley JK: Benzodiazepine receptor binding deficit in audiogenic seizure susceptible mice. Fed Proc 44: 1106, 1985.

Olsen RW, Wamsley JK, Lee RJ, Lomax P: Benzodiazepine/barbiturate/GABA receptor-chloride ionophore complex in a genetic model for generalised epilepsy. In Delgado-Escueta AV, Ward AA, Woodbury DM, Porter RJ (eds): "Advances in Neurology." New York: Raven Press, 44: 1986, 365–378.

Olsen RW, Wamsley JK, Lee R, Lomax P: Alterations in the benzodiazepine/GABA receptor-chloride ion channel complex in the seizure-sensitive Mongolian Gerbil. In Farrielo RG, Morselli PL, Lloyd KG, Quesney LF, Engel G (eds): "Neurotransmitters, Seizures and Epilepsy II." New York: Raven Press, 1984, pp 201–213.

Pedley TA, Horton RW, Meldrum BS: Electroencephalographic and behavioural effects of a GABA agonist (muscimol) on photosensitive epilepsy in the baboon, Papio papio. Epilepsia 20: 409–416, 1979.

Peeters BWMM, van Rijn CM, Vossen JMH, Coenen AML: Effects of GABA-ergic agents on spontaneous non-convulsive epilepsy, EEG and behaviour, in the WAG/Rij inbred strain of rats. Life Sci 45: 1171–1176, 1989.

Peterson DW, Collins JF, Bradford HF: The kindling model of epilepsy: Anticonvulsant action of amino acid antagonists. Brain Res 275: 169–172, 1983.

Peterson GM, Ribak CE, Oertel WH: A regional increase in the number of hippocampal GABAergic neurons and terminals in the seizure-sensitive gerbil. Brain Res 340: 384–389, 1985.

Przewlocki R, Lason W, Stach R, Kacz D: Opioid peptides, particularly dynorphin, after amygdaloid-kindled seizures. Regul Pept 6: 385–392, 1983.

Reigel CE, Dailey JW, Jobe PC: The genetically epilepsy-prone rat: an overview of seizure prone characteristics and responsiveness to anticonvulsant drugs. Life Sci 39: 763–774, 1986.

Ribak CE: Axon terminals of GABAergic chandelier cells are lost at epileptic foci. Brain Res 326: 251–260, 1985.

Ribak CE, Bradburne RM, Harris AB: A preferential loss of GABAergic symmetric synapses in epileptic foci: A quantitative ultrastructural analysis of monkey neocortex. J Neurosci 2: 1725–1735, 1982.

Ribak CE, Hunt CA, Bakay RAE, Oertel WH: A decrease in the number of GABAergic somata is associated with the preferential loss of GABAergic terminals at epileptic foci. Brain Res 363: 78–90, 1986.

Ribak CE, Byun MY, Ruiz GT, Reiffenstein RJ: Increased levels of amino acid neurotransmitters in the inferior colliculus of the genetically epilepsy-prone rat. Epilepsy Res 2: 9–13, 1988.

Ribak CE, Joubran C, Kesslak JP, Bakey RAE: A selective decrease in the number of GABAergic somata occurs in pre-seizing monkeys with alumina gel granuloma. Epilepsy Res 4: 126–138, 1989.

Roberts RC, Kim HL, Ribak CE: Increase numbers of neurons occur in the inferior colliculus of the young genetically epilepsy prone rat. Dev Brain Res 23: 277–281, 1985a.

Roberts RC, Ribak CE, Oertel WH: Increased numbers of GABAergic neurons occur in the inferior colliculus of an audiogenic model of genetic epilepsy. Brain Res 361: 324–338, 1985b.

Robertson HA: Audiogenic seizures: increased benzodiazepine receptor binding in a susceptible strain of mice. Eur J Pharmacol 66: 249–252, 1980.

Ross SM, Craig CR: γ-Aminobutyric acid concentration, L-glutamate 1-decarboxylase activity, and properties of the γ-aminobutyric acid postsynaptic receptor in cobalt epilepsy in the rat. J Neurosci 1: 1388–1396, 1981a.

Ross SM, Craig CR: Studies on γ-aminobutyric acid transport in cobalt experimental epilepsy in the rat. J Neurochem 36: 1006–1011, 1981b.

Savage DD, Nadler JV, McNamara JO: Reduced kainic acid binding in rat hippocampal formation after limbic kindling. Brain Res 323: 128–131, 1984a.

Savage DD, Werling LL, Nadler JV, McNamara JO: Selective and reversible increase in the number of quisqualate-sensitive glutamate binding sites on hippocampal synaptic membranes after angular bundle kindling. Brain Res 307: 332–335, 1984b.

Schechter PJ, Tranier Y, Yung MJ, Böhlen P: Audiogenic seizure protection by elevated GABA concentration in mice: effects of γ-acetylenic GABA and γ-vinyl GABA, two irreversible GABA-T inhibitors. Eur J Pharmacol 45: 319–328, 1977.

Schlesinger K, Boggan W, Freedman DX: Genetics of audiogenic seizures: 1. Relation to brain serotonin and norepinephrine in mice. Life Sci 4: 2345–2351, 1965.

Shin C, Pedersen HB, McNamara JO (1985): γ-Aminobutyric acid and benzodiazepine receptors in the kindling model of epilepsy: A quantitative radiohistochemical study. J Neurosci 5: 2696–2701.

Shirasaka Y, Ito M, Tsuda H, Shiraishi H, Oguro K, Mutoh K, Mikawa H: Quantitative autoradiographic study of benzodiazepine receptors: Comparative study between normal ddY mouse and El mouse. Jap J Psychiat Neurol 43: 563–564, 1989.

Snead OC: Gamma hydroxybutyric acid, gamma aminobutyric acid, and petit mal epilepsy. In Fariello RG, Morselli PL, Lloyd KG, Quesney LF, Engel J (eds): "Neurotransmitters, Seizures and Epilepsy II. "New York: Raven Press, 1984, pp 37–47.

Spyrou NA, Prestwich SA, Horton RW: Synaptosomal [³H] GABA uptake and [³H] nipecotic acid binding in audiogenic seizure susceptible (DBA/2) and resistant (C57BL/6) mice. Eur J Pharmacol 100: 207–210, 1984.

Squires R, Naquet R, Riche D, Braestrup C: Increased thermolability of benzodiazepine receptors in cerebral cortex of a baboon with spontaneous seizures: A case report. Epilepsia 20: 215–221, 1979.

Suzuki J, Nakamoto Y, Shinkawa Y: Local cerebral glucose utilization in epileptic seizures of the mutant El mouse. Brain Res 266: 359–363, 1983.

Sykes CC, Horton RW: Cerebral glutamic acid decarboxylase activity and γ-aminobutyric acid concentrations in mice susceptible or resistant to audiogenic seizures. J Neurochem 39: 1489–1491, 1982.

Tacke U, Braestrup C: A study of benzodiazepine receptor binding in audiogenic seizure-susceptible rats. Acta Pharmacol Toxicol 55: 252–259, 1984.

Tacke U, Tuomisto L, Danner R: Cortical spike wave discharges during audiogenic seizures in rats. Exp Neurol 85: 233–238, 1984.

Ticku MK: Differences in γ-aminobutyric acid receptor sensitivity in inbred strains of mice. J Neurochem 32: 1135–1138, 1979.

Trottier S, Berger B, Chauvel P, Dedek J, Gay M: Alterations of the cortical noradrenergic system in chronic cobalt epiloptogenic foci in the rat: A histofluorescent and biochemical study. Neuroscience 6: 1069–1080, 1981.

Trottier S, Claustre Y, Caboche J, Dedek J, Chauvel P, Nassif S, Scatton B (1983): Alterations of noradrenaline and serotonin uptake and metabolism in chronic cobalt-induced epilepsy in the rat. Brain Res 272: 255–262.

Tuff LP, Racine RJ, Mishra RK: The effects of kindling on GABA-mediated inhibition in the dentate gyrus of the rat. II Receptor binding. Brain Res 277: 91–98, 1983.

Tunnicliff G, Wimer CC, Wimer RE: Relationships between neurotransmitter metabolism and behaviour in seven inbred strains of mice. Brain Res 61: 428–434, 1973.

Vergnes M, Marescaux C, Micheletti G, Reis J, Depaulis A, Rumbach L, Warter JM: Spontaneous paroxysmal electroclinical non-convulsive epilepsy. Neurosci Lett 33: 97–101, 1982.

Vergnes M, Marescaux C, Micheletti G, Depaulis A, Rumbach L, Warter JM: Enhancement of spike and wave discharges by GABA mimetic drugs in rats with spontaneous petit-mal epilepsy. Neurosci Lett 44: 91–94, 1984.

Vergnes M, Marescaux C, Depaulis A, Micheletti G, Warter JM: Ontogeny of spontaneous petit mal like seizure in Wistar rats. Dev Brain Res 30: 85–87, 1986.

Vergnes M, Marescaux C, Depaulis A, Micheletti G, Warter JM: Spontaneous spike and wave discharge in thalamus and cortex in a rat model of genetic petit mal-like seizures. Exp Neurol 96: 127–136, 1987.

Wada JA, Balzano E, Meldrum BS, Naquet R: Behavioural and electrographic effects of L-5-hydroxytryptophan and D, L parachlorophenylalanine on epileptic Senegalese baboon (*Papio papio*). Electroencephalogr Clin Neurophysiol 33: 520–526, 1972.

Willott JF, Lu SM: Midbrain pathways of audiogenic seizures in DBA/2 mice. Exp Neurol 70: 288–299, 1980.

Yoshiya L, Murashima MD, Suzuki J: The GABAergic role of epileptogenesis in the El mouse. Japn J Psychiat Neurol 43: 557–558, 1989.

7
Mechanism of GABA Agonists and Modulators of GABAergic Transmission in Anticonvulsant Activity

MAHARAJ K. TICKU

Department of Pharmacology, The University of Texas Health Science Center at San Antonio, San Antonio, TX 78284–7764

INTRODUCTION

Overall CNS excitability is maintained as a result of a balance between neurotransmitters that produce inhibitory and excitatory effects. The major inhibitory neurotransmitters include GABA and glycine, while excitatory effects are primarily mediated by different types of L-glutamate receptors. There are apparently two types of GABA receptors which differ in terms of their selectivity towards agonists and antagonists and coupling to ion channel/second messenger systems. Thus, $GABA_A$ receptors are activated by agonists like GABA and muscimol, their effects are blocked by bicuculline and picrotoxin, and facilitated by benzodiazepines and barbiturates (Olsen and Venter, 1986). $GABA_A$ receptor activation leads to the opening of specific Cl^--channels, which mediates the fast inhibitory postsynaptic potential (IPSP) response. In contrast, $GABA_B$ receptors are activated by (-) baclofen [4,amino-3(p-chlorophenyl)-butyric acid] and GABA, whose actions are blocked by phaclofen and CGP35 348 (3-aminopropyl-diäthoxymethyl-phosphinsäure). These receptors are coupled presynaptically to voltage-gated Ca^{2+}-channels, and postsynaptically they activate K^+-channels, which results in the slow IPSP response (Bowery et al., 1984; Dolphin and Scott, 1986; Dutar and Nicoll, 1988). Both pre- and postsynaptic $GABA_B$ receptors appear to be coupled to G-proteins.

In recent years, molecular biological techniques have been used to characterize $GABA_A$ receptors in the CNS. These studies have demonstrated a complex heterogeneity of $GABA_A$ receptors. Initial studies indicated the presence of only two proteins with molecular weights of ~ 51,000 (α-subunit) and ~ 56,000 (β-subunit). By photoaffinity labelling studies, the α-subunit was shown to bind [3H]benzodiazepine agonists like flunitrazepam, and the β-subunit labeled [3H]muscimol (Sigel et al., 1983; Sieghart, 1985; Stauber et al., 1987). The sequence of these two subunits exhibited sequence homology to the subunits of the well characterized nicotinic acetylcholine receptor. Based on homology, and the fact that both α- and β-subunits contained four transmembrane spanning domains (in contrast to seven such domains associated with G-protein coupled receptors), initial cloning and expression studies suggested that $\alpha_2\beta_2$ configuration was necessary to form the GABA-gated chloride channel (Schofield et al., 1987). Recent studies indicate that in addition to α- and β-subunits, another subunit, γ_2, is necessary to elicit benzodiazepine modulation of $GABA_A$ responses (Pritchett et al., 1989). Furthermore, recent cloning studies have characterized multiple variants of these subunits, including 6α-, 3β-, 2γ-, and a δ-subunit (e.g., Pritchett et al., 1989; Shivers et al., 1989). The distribution and co-localization of these subunits varies in different brain regions, and may represent physiological and pharmacological diversity of $GABA_A$ receptors. Nonetheless, a major question regarding whether a tetrameric (2α, β, γ_2) or a pentameric (2α, 2β, γ_2) stoichiometry is needed to generate functional $GABA_A$ receptors which resemble the in vivo situation is yet to be resolved. Since $GABA_A$ receptor exhibits similarities to well characterized nicotinic acetylcholine receptor, it is likely that a pentameric composition may be the preferred stoichiometry.

GABA-BENZODIAZEPINE RECEPTOR-IONOPHORE COMPLEX

Before the cloning studies, most of the information regarding the $GABA_A$ receptor system was obtained by radioligand binding and functional studies. Several laboratories using different radioligand probes demonstrated the presence of distinct binding sites on the oligomeric GABA receptor. These included GABA recognition sites (Enna, 1983), benzodiazepine/β-carboline sites (Braestrup and Squires, 1977; Möhler and Okada, 1977), picrotoxin/cage convulsant, t-butylbicyclophosphorothionate (TBPS) sites (Ticku et al., 1978; Ticku and Olsen, 1978; Squires et al., 1983; Ramanjaneyulu and Ticku, 1984; Ticku, 1986), and barbiturate sites (Maksay and Ticku, 1985; Trifilletti et al., 1984). These sites on the oligomeric receptor complex are coupled and bear an allosteric relationship to each other (Fig. 1).

The $GABA_A$ receptor site recognizes GABA agonists like GABA, muscimol, THIP (4,5,6,7-tetrahydroisoxazole[5,4-c]pyridine-3–01), isoguvacine, piperidine-4-sulfonic acid, and competitive antagonists like bicuculline. Radioligand binding studies have demonstrated the presence of high- ($K_D \sim 10$ nM) and low-affinity ($K_D \sim 100\text{-}300$ nM) sites (Enna, 1983; Olsen, 1982). Additionally, the presence of a super-low-affinity ($K_D \sim 1\mu M$) has been suggested (Olsen, 1982). The high- and low-affinity GABA receptor sites can be selectively eliminated by thiocyanate and p-diazobenzenesulfonic acid, respectively (Browner et al., 1981; Burch et al., 1983; Maksay and Ticku, 1984). Thus, the topography of the high-and low-affinity GABA receptor sites does not appear to be identical. The ability of GABA agonists to inhibit [^{3}H]GABA binding to its receptor sites correlates reasonably well with their ability to activate $GABA_A$ receptor-gated Cl^--channels.

BZ/GABA RECEPTOR IONOPHORE COMPLEX

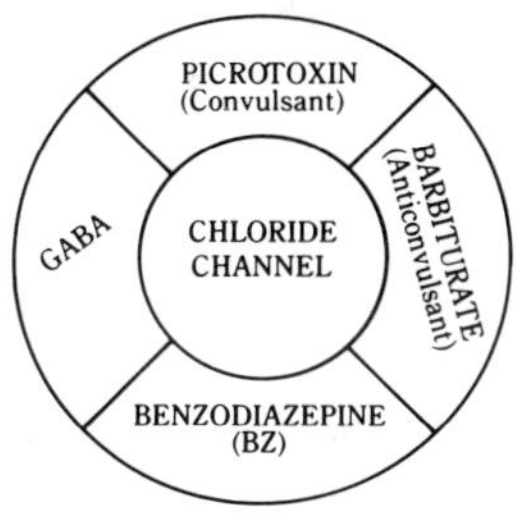

SITES OF DRUG ACTION

GABA RECEPTORS	BZ RECEPTORS	PICROTOXIN SITE	BARBITURATE SITE	COUPLING MODULATOR(S)
GABA	BZ Agonists	Picrotoxin	Barbiturates	Ethanol
Muscimol	BZ Antagonists	TBPS	(+)Etomidate	
Bicuculline	R015-4513 (?)	PTZ	Pyrazolopyridines	
R-5135 (?)	β-Carbolines (?)	R05-3663	Cyclopyrazolones	

Fig. 1. Schematic representation of $GABA_A$-benzodiazepine receptor-ionophore complex, depicting potential sites of drug modulation.

GABA AGONISTS AS ANTICONVULSANTS

The $GABA_A$ receptor system has been implicated in the pathophysiology of epilepsy (e.g., Meldrum, 1975). The idea of the involvement of the GABA receptor system in epilepsy is primarily derived from the observations that inhibitors of $GABA_A$ergic transmission produces seizures, while drugs that facilitate this system exhibit anticonvulsant effects. If there exists a deficit in the $GABA_A$ergic synapse in some form(s) of epilepsy, then several approaches could be used to enhance $GABA_A$ergic transmission. These approaches include employing $GABA_A$ agonists or using prodrugs, inhibitors of neuronal and glial GABA-uptake systems (see Schousboe, this volume), inhibitors of the enzyme(s) that metabolize GABA (see Tunnicliff, this volume), or drugs that act at other sites (e.g., benzodiazepines, barbiturates) of the GABA receptor complex and facilitate GABAergic transmission (Fig. 1). The rationale for using $GABA_A$ agonists in epilepsy would be similar to the one used for the treatment of Parkinson's disease in which the dopamine system is facilitated. However, GABA itself is a highly hydrophilic molecule and does not pass the blood-brain barrier. The topography of the GABA receptor (recognition) site also puts limitations on the size and bulk that could be attached to GABA-like molecules to make them more lipophilic. Baclofen was initially synthesized to pass the blood-brain barrier and mimic GABA-response at the $GABA_A$ receptor. However, as it turned out, baclofen does not interact with the $GABA_A$ receptor sites but acts at a distinct $GABA_B$ receptor site. Other $GABA_A$ agonists which have been tested as potential anticonvulsants include muscimol, THIP and prodrugs, progabide, and cetyl GABA (Löscher, 1982). While these GABAmimetics have been reported to be effective as anticonvulsants in some seizure models (e.g., pentylenetetrazol (PTZ), isoniazid, electroshock), they differed in their duration of action and in producing side effects like induction of loss of righting reflex (Löscher, 1982; Christensen and Krogsgaard-Larsen, 1984). Progabide was much less potent against electroshock seizures, but exerted the longest duration of action (Löscher, 1982). THIP had the shortest duration of action (1 hr), followed by muscimol (1.5–2 hr), against PTZ-induced seizures. THIP was also effective against seizures produced by picrotoxin, bicuculline, and strychnine (Mehta and Ticku, 1988). However, THIP has been reported to be ineffective against electroshock seizures (Worms and Lloyd, 1981; Löscher, 1982).

One of the problems associated with using GABA$_A$ agonists and indeed any GABAmimetic relates to the ubiquitous distribution of GABA receptors in the CNS. While it is likely that in some form(s) of epilepsy there may exist a localized deficit in GABA$_A$ergic transmission, GABA$_A$ agonists will act on all the GABA receptors, resulting in CNS effects that are unwarranted. Since GABA$_A$ receptors are ubiquitously distributed, and have been implicated in motor, sensory, psychic and endocrine functions, unwarranted side effects will be difficult to control with GABA agonists. In fact, in experimental studies, GABAmimetics have been reported to produce effects like general sedation, decreased motor coordination and hunched posture (Löscher, 1980; Palfreyman et al., 1981), effects which are related to enhanced GABAergic tone. Furthermore, some direct acting GABA agonists like muscimol and THIP have been reported to produce epileptogenic action (e.g., Golden and Fariello, 1984). Other problems associated with a GABA replenishment approach have been discussed elsewhere (Fariello and Ticku, 1983). Thus, while some currently available GABA$_A$ agonists are effective anticonvulsants in several seizure models, their use is limited by their inability to pass the blood-brain barrier, short duration of action, and associated toxicity.

BARBITURATES

Barbiturates like phenobarbital and mephobarbital have been used effectively as anticonvulsants for decades. These barbiturates are primarily effective against partial and generalized tonic-clonic seizures (Booker, 1982; Rall and Schleifer, 1985). Barbiturates produce their CNS depressant effects in a concentration-dependent manner; lower doses producing muscle relaxant/anticonvulsant effects, while higher doses produce anesthetic effects. There are several lines of evidence which implicate the GABA$_A$ergic system in the pharmacological actions of barbiturates.

One of the earliest indications that barbiturates may act via GABA$_A$ receptor system came following the identification and characterization of the picrotoxin site with [^{3}H] α-dihydropicrotoxinin (DHP). The site labeled with [^{3}H] α-DHP was distinct from the GABA receptor site, and its binding was inhibited by picrotoxin-like convulsants and barbiturates (Ticku et al.,

1978). This led to the suggestion that barbiturates may act at the picrotoxin site (Ticku and Olsen, 1978). Further characterization of the picrotoxin site with a superior radioligand, [^{35}S]t-butylbicyclophosphorothionate (TBPS; TBPT), confirmed that this was an important site of action, since a variety of convulsants, and several anxiolytics and barbiturates, inhibited the binding of TBPS (Squires et al., 1983; Ramanjaneyulu and Ticku, 1984; Ticku, 1986). It was also shown that barbiturates inhibited the [^{35}S] TBPS binding noncompetitively and accelerated the disassociation profile of [^{35}S] TBPS, relative to competitive convulsant ligands (Ticku and Maksay, 1983; Ramanjaneyulu and Ticku, 1984; Trifiletti et al., 1984; Maksay and Ticku, 1985). Figure 2 shows that anticonvulsant barbiturate, R(-)N-methyl-5-phenylpropyl barbituric acid (MPPB), and pentobarbital greatly accelerate the dissociation of [^{35}S]TBPS from rat brain membranes. This led to the suggestion of a distinct "barbiturate" receptor site on the oligomeric GABA receptor complex (Ticku and Maksay, 1983; Maksay and Ticku, 1985).

Further support to the GABA hypothesis of barbiturate action is provided by the observations that barbiturates enhance the specific binding of [^{3}H] benzodiazepine and [^{3}H]GABA agonists (Olsen, 1982). Interestingly, phenobarbital does not enhance the equilibrium binding of either [^{3}H] diazepam or [^{3}H]GABA, however, it does block the enhancing effect of pentobarbital, suggesting a mutually exclusive site (Leeb-Lundberg et al., 1980; Leeb-

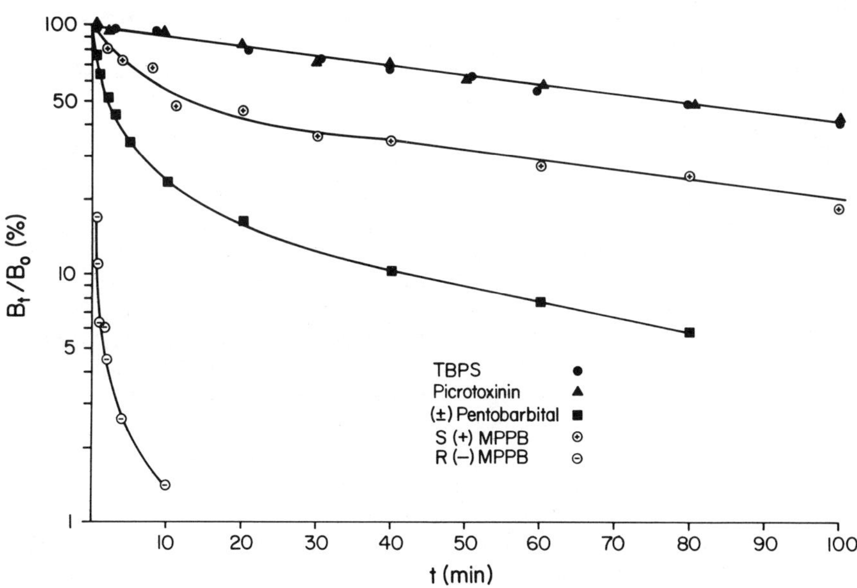

Fig. 2. Dissociation of [^{35}S]TBPS binding from brain membranes by convulsant and anticonvulsant drugs (modified from Maksay and Ticku, 1985).

Lundberg and Olsen, 1982; Skolnick et al., 1980; Ticku, 1981; Olsen, 1982; Thyagarajan et al., 1983). In contrast, both phenobarbital and pentobarbital inhibit picrotoxin/TBPS binding (Ticku and Olsen, 1978; Ramanjaneyulu and Ticku, 1984). These binding studies tend to suggest that the interaction of the anticonvulsant phenobarbital with the picrotoxin site may be important for its therapeutic effect as an anticonvulsant. Some support to this hypothesis is provided by the behavioral observations that phenobarbital was much less effective against bicuculline- vs picrotoxin-induced seizures (Tables I and II; Mehta and Ticku, 1988). This is also consistent with our previous observation that bicuculline prevented the anticonvulsant effect of pentobarbital, but not that of phenobarbital, against electroshock seizures (Rastogi and Ticku, 1986). These results agree with the electrophysiological study of Simmonds (1981), who has shown that phenobarbital reduced the potency of picrotoxin, but not of bicuculline, as GABA antagonists. These observations appear to be consistent with the binding studies where phenobarbital does not interact with the GABA binding sites, while it inhibits picrotoxin binding. Furthermore, pentobarbital, but not phenobarbital, potentiated the anticonvulsant action of diazepam against chemoconvulsants (Tables I and II; Mehta and Ticku, 1988). This observation is consistent with binding studies where pentobarbital, but not phenobarbital, enhance diazepam binding. In summary, radioligand binding and behavioral studies tend to suggest that the interaction of phenobarbital with the picrotoxin site may be crucial for its anticonvulsant effect. In contrast, pentobarbital's interactions with picrotoxin, GABA, and benzodiazepine sites may be important for its anticonvulsant and hypnotic effects.

Electrophysiological data support the notion that barbiturates enhance both the presynaptic and postsynaptic inhibition mediated by GABA. Barbiturates abolish the paroxysmal depolarizing events induced by GABA antagonist bicuculline in spinal cord neurons (Macdonald and Barker, 1978a; 1979). Both phenobarbital and pentobarbital enhance GABA-mediated responses at lower concentrations and activate GABA-gated Cl^--channels directly at higher concentrations , with pentobarbital being more potent than phenobarbital (Nicoll, 1975; Nicoll and Wojtowicz, 1980; Simmonds, 1981; Barker and Ransom, 1978; Macdonald and Barker, 1978a, 1979). In the cuneate nucleus, barbiturates reduce the potency of picrotoxin as a muscimol antagonist (Simmonds, 1981). A recent study by Macdonald et al. (1989) has demonstrated that barbiturates like phenobarbital potentiate the GABA receptor current by increasing mean channel time and prolonging bursts, without altering burst frequency. These investigators have characterized GABA-induced bursts into short duration (B_1 and B_2) and long burst state (B_3). They have postulated that barbiturate action appears to increase the relative frequency of long (B_3) bursts and decrease the proportion of shorter (B_1, B_2) bursts. Finally, it is not known which of the subunits that encode GABA$_A$ receptor binds barbiturates, since α- or β-subunits alone, or in combination ($\alpha_2\beta_2$), has been shown to express GABA-gated Cl^--channels, and the response of GABA was enhanced by barbiturates and blocked by picrotoxin

TABLE I. Effect of Various Drugs on Bicuculline (8 mg/kg, IP)-Induced Convulsions in Rats

Drug	Dose (mg/kg, IP)	n	Onset of convulsion (min) mean ± S.D.	Duration of seizures (min) mean ± S.D. Clonic	Tonic	Mortality following administration of convulsant (min) mean ± S.D.	Mortality (/N)
Control	—	11	1.2 ± 0.48	0.6 ± 0.33	7.6 ± 2.05	11.7 ± 3.84	11/11
Pentobarbital	10	8	2.1 ± 0.79[*]	1.0 ± 0.09	7.8 ± 1.58	30.6 ± 2.61[**]	5/8
	20	7	5.0 ± 1.00[**]	1.9 ± 2.04	1.8 ± 1.95[**]	65.0 ± 7.07[**]	2/7[**]
Phenobarbital	45	9	1.0 ± 0.20	1.2 ± 0.23	11.1 ± 1.85	16.2 ± 2.54	9/9
THIP	15	6	1.2 ± 0.42	0.7 ± 0.11	8.1 ± 1.42	15.0 ± 3.22	6/6
	30	6	2.6 ± 1.02[**]	0.6 ± 0.18	2.1 ± 0.45[**]	24.3 ± 14.01[*]	3/6[*]
Baclofen	20	6	0.8 ± 0.25	0.5 ± 0.19	8.0 ± 1.69	13.3 ± 3.08	6/6
Diazepam	0.5	6	1.0 ± 0.44	0.5 ± 0.24	8.9 ± 1.02	15.0 ± 4.20	6/6
	2	6	3.3 ± 0.44[**]	0.7 ± 0.26	4.4 ± 1.36[**]	25.7 ± 12.50[**]	3/6[*]
THIP 15 + pentobarbital 10		6	2.9 ± 0.66[**]	0.5 ± 0.08	3.5 ± 0.89[**]	24.5 ± 7.78[*]	2/6[*]
Baclofen 10 + pentobarbital 10		6	1.6 ± 0.49	0.6 ± 0.23	3.0 ± 0.73[**]	26.0 ± 5.66[**]	2/6[*]
Pentobarbital 10 + diazepam 0.5		6	3.0 ± 0.64[**]	0.6 ± 0.41	2.5 ± 1.24[**]	40.0 ± 0.00[**]	1/6[**]
THIP 15 +phenobarbital 30		6	1.0 ± 0.26	0.5 ± 0.09	8.7 ± 0.88	17.8 ± 2.79	6/6
Baclofen 10 +phenobarbital 30		6	0.8 ± 0.26	0.6 ± 0.18	8.2 ± 1.38	17.7 ± 3.01	6/6
Phenobarbital 30 +diazepam 0.5		6	1.0 ± 0.18	0.6 ± 0.27	10.5 ± 3.86	16.7 ± 2.16	6/6

[*]P < 0.05; [**]P <0.005 as compared to control. Reproduced with permission from Mehta and Ticku (1988).

TABLE II. Effect of Various Drugs on Picrotoxin (10 mg/kg, IP)-Induced Convulsions in Rats

Drug	Dose (mg/kg, IP)	n	Onset of convulsion (min) mean ± S.D.	Duration of seizures (min) mean ± S.D. Clonic	Tonic	Mortality following administration of convulsant (min) mean ± S.D.	Mortality (/N)
Control	—	16	6.0 ± 1.05	0.7 ± 0.28	12.7 ± 3.63	23.9 ± 6.02	16/16
Pentobarbital	5	6	6.4 ± 1.02	0.8 ± 0.45	13.4 ± 2.95	28.2 ± 5.71	6/6
	20	6	12.3 ± 2.16[**]	0.8 ± 0.16	0.3 ± 0.51[**]	—	0/6[**]
Phenobarbital	10	8	6.2 ± 0.71	1.1 ± 0.14	14.3 ± 3.35	32.0 ± 5.10	6/8
	30	8	9.0 ± 0.76	1.3 ± 0.77	2.2 ± 0.82[**]	30.0 ± 0.00	2/8[**]
THIP	15	6	6.1 ± 0.80	0.7 ± 0.09	14.7 ± 1.90	25.3 ± 3.78	6/6
	30	6	14.7 ± 3.50[**]	0.5 ± 0.11	6.2 ± 0.89[**]	40.5 ± 13.70[**]	4/6
Baclofen	20	6	6.6 ± 1.36	0.6 ± 0.19	12.8 ± 4.52	31.0 ± 12.84	6/6
Diazepam	0.5	6	7.5 ± 1.52	0.7 ± 0.28	12.1 ± 1.70	23.2 ± 3.97	6/6
	2	6	15.8 ± 1.47[**]	0.9 ± 0.37	0.7 ± 0.60[**]	—[**]	0/6[**]
THIP	15						
+pentobarbital	5	6	13.8 ± 2.79[**]	0.8 ± 0.45	5.3 ± 1.69[**]	52.5 ± 10.61[**]	2/6[**]
Baclofen	10						
+pentobarbital	5	6	14.0 ± 3.16[**]	0.6 ± 0.16	4.3 ± 1.28[**]	45.0 ± 0.00[**]	1/6[**]
Pentobarbital	5						
+diazepam	0.5	6	11.7 ± 1.91[**]	0.8 ± 0.15	1.6 ± 1.44[**]	—[**]	0/6[**]
THIP	15						
+phenobarbital	10	6	16.5 ± 1.87[**]	0.5 ± 0.17	2.9 ± 0.83[**]	60.0 ± 15.00[**]	3/6[*]
Baclofen	10						
+phenobarbital	10	6	8.0 ± 2.28	0.8 ± 0.24	3.6 ± 1.67[**]	30.0 ± 7.07	2/6[**]
Phenobarbital	10						
+diazepam	0.5	6	6.4 ± 0.53	1.0 ± 0.06	13.8 ± 2.64	36.8 ± 3.95[**]	4/6

[*]P < 0.05; [**]P < 0.005 as compared to control; (-) indicates no mortality. Reproduced with permission from Mehta and Ticku (1988).

(Schofield et al., 1987). The γ_2-subunit which is crucial to demonstrate benzodiazepine enhancement of $GABA_A$ responses is apparently not needed for barbiturate action.

Besides enhancing $GABA_A$-mediated inhibition, barbiturates also decrease glutamate (non-NMDA)-mediated excitatory transmission (Macdonald and Barker, 1978a), and inhibit voltage-gated Ca^{2+}-channels (Werz and Macdonald, 1985; Twombly et al., 1987). Since both of these systems are involved in the inhibition of neurotransmitter release, a potential role of these systems in the anticonvulsant and/or hypnotic effects of barbiturates cannot be ruled out.

BENZODIAZEPINES

Another class of anticonvulsant drugs that appear to produce their therapeutic effect via the $GABA_A$ receptor system is the benzodiazepines. These drugs include diazepam and clonazepam, which are effective against several forms of epilepsy, including status-epilepticus and primary generalized and partial seizures. Several lines of evidence, including radioligand binding studies, electrophysiological and behavioral, lend support to the GABAergic mechanism for benzodiazepine actions, including anticonvulsant effects.

An earlier report by Schmidt et al. (1967) that benzodiazepines facilitated the presynaptic inhibition received little attention for years, since presynaptic inhibition was not a widely accepted concept by physiologists and pharmacologists. However, subsequent studies by two groups implicated the GABA receptor system in the action of benzodiazepines (Polc et al., 1974; Costa et al., 1975). Diazepam was shown to enhance GABA-mediated pre- and postsynaptic inhibition in cuneate nucleus (Polc et al., 1974; Polc and Haefely, 1976). Furthermore, prior treatment with inhibitors of GABA synthesis eliminated diazepam's enhancing effect, thereby implicating a role for ongoing GABA tone in the actions of benzodiazepines (Polc et al., 1974, Polc and Haefely, 1976). This was followed by the discovery of the high-affinity benzodiazepine receptor sites in 1977 (Braestrup and Squires, 1977; Möhler and Okada, 1977), and demonstration of coupling of these receptor sites to the GABA receptor (Tallman et al., 1980). The high-affinity benzodiazepine receptor shows specificity for a series of benzodiazepines and beta-carbolines, which correlated well with their potency as anticonvulsants against chemoconvulsions produced by GABA antagonists (Duka et al., 1979; Paul et al., 1979; Meldrum and Braestrup, 1984). There is also evidence for heterogeneity of benzodiazepine binding sites (Squires et al., 1979; Young et al 1981.). The benzodiazepine receptor site on the oligomeric GABA receptor complex is a unique site, since it can be modulated by drugs with diverse and opposite pharmacological properties, with positive and negative intrinsic efficacies (Braestrup et al., 1984; Nutt, 1986; Little, 1988). Thus, full agonists like diazepam produce anxiolytic/anticonvulsant effects, whereas inverse agonists like betacarbolines produce anxiogenic and in some cases, convulsant effects. The effects of both agonists and inverse agonists are mediated via the same receptor site, since they can be blocked by the neutral ligand, flumazenil (Ro 15–1788). Based on our current understanding of behavioral pharmacology of drugs that modulate the benzodiazepine binding site, it is likely that some new drugs that are partial agonists at this site may prove to be highly effective anticonvulsants, and may lack the unwanted sedative effect.

The possibility also exists that drugs will be discovered in the near future which could modulate the picrotoxin and barbiturate sites in a manner similar to the benzodiazepine site. Thus, partial agonists at the picrotoxin site could prove to be effective anticonvulsant drugs.

In biochemical assays, benzodiazepine receptor agonists increase GABA-induced ^{36}Cl-influx, whereas inverse agonists decrease it (Mehta and Ticku, 1989a). Both the enhancing effect of agonists and the inhibitory effect of inverse agonists can be reversed by flumazenil. Benzodiazepines do not alter the GABA function following the occupation of all GABA receptors with the agonist (Mehta and Ticku, 1989a). This observation is consistent with the notion that benzodiazepines need ongoing GABAergic tone to produce their effects (Polc et al., 1974). Thus, the effects of benzodiazepine agonists are very subtle, relative to barbiturates. Electrophysiological studies have demonstrated that benzodiazepine receptor agonists increase the frequency of opening of GABA$_A$-gated Cl$^-$-channels (Costa et al., 1975; Choi et al., 1977; Macdonald and Barker, 1978b; Chan and Farb, 1985). Cloning studies have indicated that while α-subunits of the GABA receptor bind [^{3}H]flunitrazepam by photoaffinity labeling, for benzodiazepine enhancement of GABA responses, a combination of α-, β-, and γ2-subunits is needed (Pritchett et al., 1989). Although the γ2-subunit is not photoaffinity labeled by [^{3}H]flunitrazepam, it is crucial for demonstrating the functional response to benzodiazepines. It could be speculated that the γ2-subunit is necessary for producing the conformation change following benzodiazepine binding to the α-subunit that results in the facilitation of the GABA responses.

One of the major problems with the benzodiazepines is the development of pharmacodynamic tolerance to their anticonvulsant effect (Rosenberg et al., 1989; File, 1984; Gallager et al., 1984; Little, 1988). Furthermore, there also exists cross tolerance between benzodiazepines and other drugs that facilitate GABA$_A$ergic transmission, e.g., ethanol and barbiturates (Little, 1988). The exact molecular mechanism(s) involved in tolerance to the anticonvulsant effect of benzodiazepine agonists has yet to be defined. In electrophysiological experiments, chronic benzodiazepine agonist treatment produces subsensitivity to GABA and benzodiazepines (Gallagher et al., 1984). In behavioral studies, chronic benzodiazepine agonist treatment increases the efficacy of inverse agonists (Little, 1988). These results suggest a shift in the conformation of the receptor complex towards a state which prefers inverse agonists. Radioligand binding studies have provided contradictory results, with some reporting down-regulation (Rosenberg et al., 1989), others reporting no changes in the number of benzodiazepine receptors (Gallager et al., 1984; Braestrup et al., 1979). Interestingly, we have observed that the binding (Mhatre and Ticku, 1989) following chronic ethanol treatment, and efficacy following ethanol withdrawal of inverse agonists as ethanol antagonists, is also increased (Mehta and Ticku, 1989b). Thus, possible similar molecular mechanisms may exist for the development of tolerance/withdrawal to drugs which facilitate $GABA_A$ergic transmission like benzodiazepines, ethanol, and barbiturates.

ROLE OF $GABA_A$ AND $GABA_B$ RECEPTORS

While the role of $GABA_A$ receptors in epilepsy and in the anticonvulsant effect of several classes of drugs is well recognized, there is no evidence for the involvement of $GABA_B$ receptors in either seizure disorders or anticonvulsant action. Furthermore, $GABA_A$ receptors are more abundant than $GABA_B$ receptors in the CNS. It may also be noted that all known $GABA_A$ antagonists produce seizures in all animal models, whereas $GABA_B$ antagonists like phaclofen, and the newly discovered antagonist, CGP35 348, do not produce seizures in animals. These findings could be interpreted to imply that $GABA_B$ receptors do not have a profound role in regulating overall CNS excitability. Furthermore, the $GABA_B$ agonist baclofen does not exhibit an anticonvulsant effect in any of the experimental models tested (Mehta and Ticku, 1986; 1988). However, I would like to suggest an adjunct role of $GABA_B$ agonists in the treatment of certain types of epilepsies. The rationale for this is that activation of presynaptic $GABA_B$ receptors usually results in decreased release of neurotransmitters, especially excitatory neurotransmitters like glutamate (e.g., Johnston, et al., 1980). Thus, a combination of a facilitator of $GABA_A$ergic transmission (i.e., enhanced postsynaptic inhibition) and a $GABA_B$ receptor agonist (i.e., decreased excitatory transmission) could result in a beneficial therapeutic effect. One of the major advantages of this approach would be to use lower doses of drugs which facilitate $GABA_A$ergic transmission (e.g., benzodiazepines, barbiturates), which in turn will result in reduced side effects of these drugs. Our laboratory has tested this possibility by using $GABA_A$ and $GABA_B$ drugs in combination at individual subeffective doses against several seizure models. Our studies have shown that activation of $GABA_B$ receptors by baclofen (up to 20 mg/kg, i.p.) alone does not produce an anticonvulsant effect against electroshock seizures or chemoconvulsions (Mehta and Ticku, 1986, 1988). Interestingly, when baclofen (10 mg/kg, i.p.) was combined with subeffective doses of pentobarbital or phenobarbital, potentiation of an anticonvulsant effect against electroshock seizures was observed (Mehta and Ticku, 1986). Likewise, baclofen (10 mg/kg, i.p.) was able to potentiate the anticonvulsant effect of subeffective dose of pentobarbital against convulsions induced by bicuculline, picrotoxin, and strychnine (Tables I and II; Mehta and Ticku, 1988). However, baclofen in combination with subeffective doses of phenobarbital was effective against picrotoxin, but not bicuculline. These differences between pentobarbital and phenobarbital may be explained by their differential interactions with various binding sites of the oligomeric $GABA_A$ receptor complex (see barbiturates above). Taken together, these results indicate that activation of GABA$_B$ receptors alone is not involved in the convulsant or anticonvulsant effect, however, their activation in combination with GABA$_A$ receptor activation tends to facilitate of the anticonvulsant effect of the barbiturates studied. These observations are of great interest and warrant further studies clinically. A major benefit could be a decrease in the incidence of side effects, which is at present, a major complication for most of the prescribed anticonvulsant drugs.

CONCLUSIONS

In summary, the use of GABA$_A$ agonists as effective anticonvulsants is not very promising. However, drugs that modulate GABA$_A$ergic transmission by other mechanisms can be effective anticonvulsants. In the future, efforts could be directed at using partial agonists of the benzodiazepine receptor site for the treatment of epilepsy. Similar approaches should be directed at identifying and/or synthesizing drugs that could act as partial agonists at picrotoxin and barbiturate sites. Partial agonists acting at these sites may provide effective anticonvulsant action with reduced severity of side effects. Another approach worth considering is the combination therapy utilizing individual subeffective doses of a GABA$_A$-mimetic and a GABA$_B$ receptor agonist such as baclofen. This drug combination appears to be effective against several seizure models. The rationale for this approach is also to use effective anticonvulsants with reduced side effects. Finally, another drug combination not discussed in this chapter includes using subeffective doses of GABA$_A$-mimetic and N-methyl-D-aspartate (NMDA) antagonists.

ACKNOWLEDGMENTS

I thank Drs. S.K. Rastogi, A.K. Mehta, and M. Mhatre for their research contribution, and Mrs. Diana Reese for excellent secretarial assistance. Supported by NIH grant NS-15339.

REFERENCES

Barker JL, Ransom BR: Pentobarbital pharmacology of mammalian central neurons grown in tissue culture. J Physiol (Lond) 280:355–372, 1978.

Booker ME: Phenobarbital: Relation of plasma concentration to seizure control. In Woodbury DM, Penry JK, Pippenger CE (eds): "Antiepileptic Drugs." New York: Raven Press, 1982, pp 341–350.

Bowery NG, Price GW, Hudson AL, Wilkin GP, Turnball MJ: GABA receptor multiplicity: visualization of different receptor types in the mammalian CNS. Neuropharmacol 23:219–231, 1984.

Braestrup C, Honoré T, Nielsen M, Petersen EN, Jensen LH: Ligands for benzodiazepine receptors with positive and negative efficacy. Biochem Pharmacol 33: 859–862, 1984.

Braestrup C, Nielsen M, Squires RF: No change in rat benzodiazepine receptors after withdrawal from continuous treatment with lorazepam and diazepam. Life Sci 24:347–350, 1979.

Braestrup C, Squires RF: Specific benzodiazepine-receptors in rat brain characterized by high affinity [^{3}H]diazepam binding. Proc Natl Acad Sci USA 74:3805–3809, 1977.

Browner M, Ferkany JW, Enna SJL: Biochemical identification of pharmacologically and functionally distinct GABA receptors in rat brain. J Neurochem 5:514–518, 1981.

Burch TP, Thyagarajan R, Ticku MK: Group selective reagent modification of the benzodiazepine-γ-aminobutyric acid receptors complex reveals that low affinity γ-aminobutyric acid receptors stimulate benzodiazepine binding. Mol Pharmacol 23:52–59, 1983.

Chan CY, Farb DM: Modulation of neurotransmitter action: control of the γ-aminobutyric acid response through the benzodiazepine receptor. J Neurosci 5:2365–2373, 1985.

Choi DW, Farb DM, Fischbach GD: Chlordiazepoxide selectively augments GABA action in spinal cord cell cultures. Nature 269:342–345, 1977.

Christensen AV, Krogsgaard-Larsen P: GABA agonists: molecular and behavioral pharmacology. In Fariello R, Morselli PL, Lloyd KG, Quesney LF, Engel J (eds): "Neurotransmitters, Seizures and Epilepsy II." Raven Press, 1984, pp 109–126.

Costa E, Guidotti A, Mao CC, Suria A: New concepts in the mechanism of action of benzodiazepines. Life Sci 7:167–186, 1975.

Dolphin AC, Scott RH: Inhibition of calcium currents in cultured rat dorsal root ganglion neurons by (-)baclofen. Br J Pharmacol. 88:213–220, 1986.

Duka T, Hollt V, Herz A: In vivo receptor occupation by benzodiazepines and correlation with the pharmacological effects in mouse. Brain Res 179:147–156, 1979.

Dutar P, Nicoll RA: A physiological role for GABA$_B$ receptors in the central nervous system. Nature 332:156–158, 1988.

Enna SJ: GABA receptors. In Enna SJ (ed): "The GABA Receptors." Clifton, NJ: Humana Press, 1988, pp 1–23.

Fariello RF, Ticku MK: The perspective of GABA replenishment therapy in the epilepsies: a critical evaluation of hopes and concerns. Life Sci 33:1629–1640, 1983.

File SE: Behavioral pharmacology of benzodiazepines. Proc Neuropsychopharm Biol Psychiatry 8:19–31, 1984.

Gallager DW, Lakoski JM, Gonsalves SF, Rauch SL: Chronic benzodiazepine treatment decreases postsynaptic GABA sensitivity. Nature 308:74–76, 1984.

Golden GT, Fariello RF: Epileptogenic action of some direct GABA agonists: effects of manipulation of the GABA and glutamate systems. In Fariello RG, Morselli PL, Lloyd KG, Quesney LF, Engel J (eds): "Neurotransmitters, Seizures and Epilepsy II." New York: Raven Press, 1984, pp 237–244.

Johnston GAR, Hailstone MH and Freeman CG: Baclofen: Stereoselective inhibition of excitant amino acid release. J Pharm Pharmacol 32:230–231, 1980.

Leeb-Lundberg F, Olsen RW: Interaction of barbiturates of various pharmacological categories with benzodiazepine receptors. Mol Pharmacol 21:320–328, 1982.

Leeb-Lundberg F, Snowman A, Olsen RW: Barbiturate receptors are coupled to benzodiazepine receptors. Proc Natl Acad Sci USA 77:7468–7472, 1980.

Little HJ: Chronic benzodiazepine treatment increases the effects of inverse agonists. In Biggio G, Costa E (eds): "Chloride Channels and Their Modulation by Neurotransmitters and Drugs." New York: Raven Press, Vol. 45, 1988, pp 307–324.

Löscher W: A comparative study of the pharmacology of inhibitors of GABA-metabolism. Naunyn Schmiedebergs Arch Pharmacol 315:119–128, 1980.

Löscher W: Comparative assay of anticonvulsant and toxic potencies of sixteen GABAmimetic drugs. Neuropharmacology 21:803–810, 1982.

Macdonald RL, Barker JL: Different actions of anticonvulsant and anesthetic barbiturates revealed by use of cultured mammalian neurons. Science 200:775–777, 1978a.

Macdonald RL, Barker JL: Benzodiazepines specifically modulate GABA-mediated postsynaptic inhibition in cultured mammalian neurons. Nature 271:563–564, 1978b.

Macdonald RL, Barker JL: Anticonvulsant and anesthetic barbiturates: different postsynaptic actions on cultured mammalian neurons. Neurology 29:432–447, 1979.

Macdonald RL, Rogers CJ, Twyman RE: Barbiturate regulation of kinetic properties of the GABA$_A$ receptor channel of mouse spinal neurons in culture. J Physiol (Lond) 417:483–500, 1989.

Maksay G, Ticku MK: Diazotization and thiocynate differentiate agonists from antagonists for the high- and low-affinity receptors. J Neurochem 43: 261–268, 1984.

Maksay G, Ticku MK: Dissociation of [^{35}S]t-butylbicyclophosphorothionate binding differentiates convulsant and depressant drugs that modulate GABAergic transmission. J Neurochem 44:480–485, 1985.

Mehta AK, Ticku MK: Comparison of anticonvulsant effect of pentobarbital and phenobarbital against seizures induced by maximal electroshock and picrotoxin in rats. Pharmacol Biochem Behav 25:1059–1065, 1986.

Mehta AK, Ticku MK: Interaction of pentobarbital and phenobarbital with GABAergic drugs against chemoconvulsants in rats. Pharmacol Biochem Behav 30:995–1000, 1988.

Mehta AK, Ticku MK: Benzodiazepine and beta-carboline interactions with GABA$_A$ receptor-gated chloride channels in mammalian cultured spinal cord neurons. J Pharmacol Exp Ther 249:418–423, 1989a.

Mehta AK, Ticku MK: Chronic ethanol treatment alters the effect of the benzodiazepine receptor inverse agonist Ro 15–4513 in rats. Brain Res 489:93–100, 1989b.

Meldrum BJ: Epilepsy and γ-aminobutyric acid-mediated inhibition. Int. Rev. Neurobiol. 17:1–36, 1975.

Meldrum B, Braestrup G: GABA and the anticonvulsant action of benzodiazepines and related drugs. In Bowery NG (ed): "Actions and Interactions of GABA and Benzodiazepines." New York: Raven Press, 1984, pp 133–139.

Mhatre M, Ticku MK: Chronic ethanol treatment selectively increases the binding of partially negative ligands for benzodiazepine binding sites in cultured spinal cord neurons. J Pharmacol Exp Ther 251:164–168, 1989.

Möhler H, Okada T: Benzodiazepine receptors: demonstration in the central nervous system. Science 198:849–851, 1977.

Nicoll RA: Pentobarbital: Action on frog motoneurons. Brain Res 96:119–123, 1975.

Nicoll RA, Wojtowicz JM: The effects of pentobarbital and related compounds on frog motoneurons. Brain Res 191:225–237, 1980.

Nutt DJ: Benzodiazepine dependence in man: Reason for anxiety. TIPS. 1:457–459, 1986.

Olsen RW: Drug interaction at the GABA receptor ionophore complex. Annu Rev Pharmacol Toxicol 22:245–277, 1982.

Olsen RW, Venter J (eds): Benzodiazepine/GABA receptors and chloride channels: Structural and functional properties. "Receptor Biochemistry and Methodology." New York: Alan R. Liss, Inc., Vol. 5, 1986.

Palfreyman MG, Schechter PJ, Buckett WR, Tell GP, Koch-Weser J: The pharmacology of GABA-transaminase inhibitors. Biochem Pharmacol 30:817–824, 1981.

Paul SM, Syapin PJ, Paugh BA, Moncada A, Skolnick P: Correlation between benzodiazepine receptor occupation and anticonvulsant effects of diazepam. Nature 281:688–690, 1979.

Polc P, Haefely W: Effects of two benzodiazepines, phenobarbitone, and baclofen on synaptic transmission in the cat cuneate nucleus. Naunyn Schmiedebergs Arch Pharmacol 294:121–131, 1976.

Polc P, Möhler H, Haefely W: The effect of diazepam on spinal cord activities: possible sites and mechanisms of action. Naunyn Schmiedebergs Arch Pharmacol 284:319–337, 1974.

Pritchett DB, Sontheimer H, Shivers BD, Ymer S, Kettenmann H, Schofield PR, Seeburg PH: Importance of a novel $GABA_A$ receptor subunit for benzodiazepine pharmacology. Nature 338:582–585, 1989.

Rall TW, Schleifer LS: Drugs effective in the therapy of epilepsies. In Gilman AG, Goodman LS, Rall TW, Murad F (eds): "The Pharmacological Basis of Therapeutics." New York: Macmillan, 7th edition, 1985, pp 446–472.

Ramanjaneyulu R, Ticku MK: Binding characteristics and interaction of depressant drugs with $[^{35}S]$t-butylbicyclophosphorothionate, a ligand that binds to the picrotoxin site. J Neurochem 42:221–229, 1984.

Rastogi SK, Ticku MK: Involvement of a GABAergic mechanism in the anticonvulsant effect of pentobarbital against maximal electroshock-induced seizures in rats. Pharmacol Biochem Behav 22:141–146, 1985.

Rastogi SK, Ticku MK: Anticonvulsant profile of drugs which facilitate GABAergic transmission on convulsions mediated by a GABAergic mechanism. Neuropharmacology 25:175–185, 1986.

Rosenberg HC, Teitz EI, Chiu TH: Tolerance to anticonvulsant effects of diazepam, clonazepam, and clobazam in amygdala-kindled rats. Epilepsia 30:276–280, 1989.

Schmidt RF, Vogel ME, Zimmermann M: Die Wirkung von Diazepam auf die präsynaptische Hemmung und an den Rückenmarksreflexen. Naunyn Schmiedebergs Arch Pharmacol 258:69–82, 1967.

Schofield PR, Darlison MG, Fujita N, Burt DR, Stephenson FA, Rodriguez H, Rhee LM, Ramachandran J, Reale V, Glencorse TA, Seeburg PH, Barnard EA: Sequence and functional expression of the $GABA_A$ receptor shows a ligand-gated receptor super family. Nature 328:221–227, 1987.

Shivers BD, Killisch I, Sprengel R, Sontheimer H, Kohler M, Schofield PR, Seeburg PH: Two novel $GABA_A$ receptor subunits exist in distinct neuronal subpopulations. Neuron 3:323–337, 1989.

Sieghart W: Benzodiazepine receptors: multiple receptors or multiple conformations. J Neural Transmission 63:191–208, 1985.

Sigel E, Stephenson FA, Mamalaki C, Barnard E: A γ-aminobutyric acid/benzodiazepine receptor complex of bovine cerebral cortex: Purification and partial characterization. J Biol Chem 258:6965–6971, 1983.

Simmonds, MA: Distinction between the effects of barbiturates, benzodiazepines and phenytoin on responses to γ-aminobutyric and antagonism by bicuculline and picrotoxin. Br J Pharmacol 73:739–747, 1981.

Skolnick P, Paul SM, Barker JL: Pentobarbital potentiates GABA-enhanced [³H]diazepam binding to benzodiazepine receptors. Eur J Pharmacol 65:125–127, 1980.

Squires RF, Benson DI, Braestrup C, Coupet J, Klepner GA, Myers V, Beer B: Some properties of brain specific benzodiazepine receptor: New evidence for multiple receptors. Pharmacol Biochem Behav 10:825–830, 1979.

Squires RF, Casida JE, Richardson M, Saederup E: [³⁵S]t-Butylbicyclophosphorothionate binds with high affinity to brain specific sites coupled to γ-aminobutyric acid$_A$ and ion recognition sites. Mol Pharmacol 23:326–336, 1983.

Stauber GB, Ransom RW, Dilber AI, Olsen RW: The γ-aminobutyric acid benzodiazepine receptor protein from rat brain: Large scale purification and preparation of antibodies. Eur J Pharmacol 167:125–133, 1987.

Tallman JF, Paul SM, Skolnick P, Gallager DW: Receptors for the age of anxiety: pharmacology of the benzodiazepines. Science 207:274–281, 1980.

Thyagarajan R, Ramanjaneyulu R, Ticku MK (1983): Enhancement of diazepam and γ-aminobutyric acid binding by (+)etomidate and pentobarbital. J Neurochem 41:575–585, 1983.

Ticku MK: Interaction of depressant, convulsant and anticonvulsant barbiturates with [³H]diazepam sites at the benzodiazepine-GABA receptor-ionophore complex. Biochem Pharmacol 30:2573–2579, 1981.

Ticku MK (1986): Convulsant binding sites on the benzodiazepine/GABA receptor. In Olsen RW, Venter C (eds): "Benzodiazepine/GABA Receptors and Chloride Channels Structural and Functional Properties." New York: Alan R. Liss, Inc., 1986, pp 195–207.

Ticku MK, Ban M, Olsen RW: Binding of [³H]α-dihydropicrotoxinin, a γ-aminobutyric acid synaptic antagonist, to rat brain membranes. Mol Pharmacol 14:391–402, 1978.

Ticku MK, Maksay G: Convulsant/depressant site of action at the allosteric benzodiazepine-GABA receptor-ionophore complex. Life Sci 33:2363–2375, 1983.

Ticku MK, Olsen RW: Interaction of barbiturates with dihydropicrotoxinin binding sites related to the GABA receptor-ionophore system. Life Sci 22:1643–1656, 1978.

Trifiletti RW, Snowman AM, Snyder SH: Barbiturate recognition site on the GABA/benzodiazepine receptor complex is distinct from the picrotoxinin/TBPS recognition site. Eur J Pharmacol 106:441–447, 1984.

Twombly DA, Herman MD, Narahashi T: Pentobarbital block of voltage-dependent calcium channels in neuroblastoma cells. Abst Soc Neurosci 13:P102, 1987.

Werz MA, Macdonald RL: Barbiturates decrease voltage-dependent calcium conductance of mouse neurons in dissociated cell culture. Mol Pharmacol 28:269–277, 1985.

Worms P, Lloyd KG: Functional alterations of GABA synapses in relation to seizures. In Morselli PL, Lloyd KG, Losher W, Meldrum BS, Reynolds EH (eds): "Neurotransmitters, Seizures and Epilepsy." New York: Raven Press, 1981, pp 37–46.

Young SW, Niehoff D, Kuhar MJ, Beer B, Lippa AS: Multiple benzodiazepine receptor localization by light microscopic radiohistochemistry. J Pharmacol Exp Ther 216:425–430, 1981.

GABA Mechanisms in Epilepsy, pages 165–187
© 1991 Wiley-Liss, Inc.

8
GABA Uptake Inhibitors as Anticonvulsants

ARNE SCHOUSBOE, ORLA M. LARSSON,
AND POVL KROGSGAARD-LARSEN
PharmaBiotec Research Center,
Departments of Biology (A.S., O.M.L) and Organic Chemistry (P. K.-L.),
Royal Danish School of Pharmacy, DK-2100 Copenhagen Ø, Denmark

INTRODUCTION

The integrative function of the central nervous system (CNS) is the result of the fine tuning and coordination of excitatory and inhibitory neurotransmission processes. The quantitatively most important neurotransmitters in these systems are glutamate and GABA mediating respectively excitatory and inhibitory signals (Curtis and Johnston, 1974; Roberts, 1979; Fagg and Foster, 1983; Schousboe, 1990). There is ample evidence to suggest that imbalances between excitatory and inhibitory processes may lead to seizure activity and it is likely that such disturbances are causative factors in epilepsy (Morselli et al., 1981; Fariello et al., 1984; Nistico et al., 1986). There is a large body of direct and indirect evidence that the efficacy of GABA mediated synapses is reduced in such disorders (Meldrum, 1975; Wood, 1975; Schousboe et al., 1983a; Ribak, 1985; Krogsgaard-Larsen et al., 1987; Schousboe, 1990). Thus, in human epilepsy there may be a reduction of the activity of the GABA synthesizing enzyme glutamate decarboxylase in the epileptic foci (Lloyd et al., 1984) and also a decrease in cerebrospinal fluid levels of GABA (Wood et al., 1979; Löscher and Siemes, 1985).

Different strategies may be followed in order to enhance the functional efficacy of the GABA mediated inhibitory processes. Synaptic release of GABA may be enhanced by stimulation of GABA synthesis and/or inhibition of GABA degradation in the presynaptic nerve ending (Gram et al., 1988). Whereas a stimulation of GABA synthesis seems difficult at the present time, reduction of GABA metabolism via inhibition of GABA transaminase has proven highly successful since specific inhibitors of this enzyme such as gamma-vinyl GABA (vigabatrin) indeed enhance synaptic GABA release (Gram et al., 1988) and are clinically useful antiepileptics (Gram, 1988). For a more detailed discussion of this aspect, the reader is referred to the chapter by Tunnicliff (this volume). GABA mediated neurotransmission may also be enhanced by direct or indirect stimulation of postsynaptic GABA receptors by either GABA receptor agonists of which a large number have become available (Krogsgaard-Larsen et al., 1988) or by ligands acting allosterically at GABA receptor-associated benzodiazepine receptors or barbiturate receptors (Olsen and Venter, 1986). Also this latter approach has proven very successful in the clinical treatment of epilepsy (Gram, 1988). Alternatively, the efficacy of GABAergic synapses may be influenced and possibly enhanced by interference with the removal of GABA (Schousboe, 1979; 1990; Krogsgaard-Larsen, 1981; Schousboe et al., 1983a,b; 1990). This removal may be brought about by the high affinity transport carriers for GABA present in presynaptic GABAergic nerve endings as well as in surrounding astroglial cells (Schousboe, 1981; 1982).

The present review shall deal with a detailed discussion of the significance of these latter processes in relation to seizure activity. Moreover, the pharmacological properties of these processes will be described and evidence will be presented that interference with these transport processes may be a relevant approach to development of drugs with anticonvulsant and potential antiepileptic activity.

HIGH AFFINITY GABA UPTAKE
Basic Characteristics

High affinity uptake processes for GABA exhibiting K_m values in the range 1–50 µM have been reported for a variety of nervous tissue preparations including brain slices, synaptosomes, bulk prepared neurons and glial cells, and cultured neurons and glia (Martin, 1976; Hertz, 1979; Schousboe, 1981; 1982; Hösli et al., 1986). Even thirty years ago it was suggested that the ability of brain slices to accumulate GABA from the incubation medium might be related to a storage mechanism for GABA (Elliott and van Gelder, 1958). Although not entirely unequivocal the concept has emerged from a large body of circumstantial experimental evidence that such transport processes are responsible for the inactivation of neurotransmitter GABA at the synaptic level (Snyder et al., 1970; Curtis and Johnston, 1970; 1974; Iversen, 1972; Curtis et al., 1976; Schousboe, 1981; Hösli et al., 1986). In this context it should, however, be mentioned that also diffusion of GABA into the postsynaptic neuron may constitute an important mechanism by which GABA can be efficiently removed from the synaptic cleft (Hydén et al., 1986).

The ability of brain cells to accumulate GABA by high affinity transport against a concentration gradient is dependent upon the cooperative coupling of GABA uptake to sodium ions (Martin, 1976; Sellström et al., 1976; Schousboe, 1981). In this context it may be interesting to note that the coupling ratio between GABA and sodium increases from one to two as a function of development and differentiation of astroglial cells as well as GABAergic neurons (Larsson et al., 1986a). This means that the efficiency of the uptake mechanism increases as a function of the maturation and differentiation of the cells. In addition to sodium, GABA uptake is also dependent upon external chloride (Kanner, 1978).

Kanner and coworkers (Kanner, 1978; Kanner et al., 1983; Keynan and Kanner, 1988) have recently characterized highly purified and reconstituted preparations of the GABA carrier which have been found to possess the above mentioned properties. Such preparations may eventually by used to prepare antibodies to the GABA carrier. This might allow a detailed mapping of high affinity GABA uptake sites in the CNS. In addition, it would become possible to investigate possible correlations between seizure activity and aberrations in these systems. At present only little information about this is available (Ross and Craig, 1981; Spyrou et al., 1984) and it is not possible to reach any definite conclusion with regard to this important question.

Cellular Distribution

With the use of primary cultures of GABAergic neurons and different types of astrocytes it has been demonstrated that both cell types possess high affinity GABA transport systems (Schousboe, 1981; 1982; Hertz and Schousboe, 1986; Schousboe et al., 1987; 1988). These findings are well in line with autoradiographic studies of labeling of different neural preparations with ^{3}H-GABA which have shown certain neurons and astrocytes in general to be labeled (Hösli et al., 1986). Based upon quantitative studies of neuronal and astrocytic high affinity uptake of GABA, Hertz and Schousboe (1987) have proposed a model according to which the majority of synaptically released GABA is taken up into presynaptic nerve endings and about 20% of the released GABA is taken up into surrounding astrocytes. In this context it should, however, be emphasized that the capacity for GABA uptake into cultured astrocytes is subject to regulation by neuronally derived factors (Drejer et al., 1983). It is therefore conceivable that the astrocytic GABA uptake could play a more prominent role under in vivo conditions. Since the fate of GABA taken up into the different cell types is completely different, the quantitative aspects of these different cellular uptake mechanisms for GABA are functionally important (Gram et al., 1988). GABA accumulated by nerve endings is likely to be reutilized as a neurotransmitter whereas GABA accumulated into astrocytes is likely to be metabolized to tricarboxylic acid cycle constituents via GABA-transaminase which has a high activity in these cells (Schousboe et al., 1977; Hertz and Schousboe, 1986). However, ^{14}C-labeled GABA accumulated into astrocytes has been reported to be oxidized to ^{14}CO$_2$ only relatively slowly (Yu and Hertz, 1983). It may, however, not always be possible to draw definite quantitative conclusions from studies of ^{14}CO$_2$ production from labeled substrate since dilution of label in intracellular metabolite pools may obscure the results (cf. Hertz et al., 1988).

Substrate Structure-Activity Relations

Detailed kinetic analyses (Table I) of the uptake of GABA and the structurally related compounds (Fig. 1) nipecotic acid, *cis*-4-OH-nipecotic acid, *cis*-3-aminocyclohexanecarboxylic acid (ACHC), 4,5,6,7-tetrahydroisoxazolo[4,5-*c*]pyridin-3-ol (THPO) and *N*-(4,4-diphenyl-3-butenyl)-

**TABLE I. Kinetic Constants for Uptake
of GABA and Its Structural Analogues Nipecotic
Acid, *cis*-4-OH-Nipecotic Acid, *cis*-3-Aminocyclohexane Carboxylic
Acid, THPO and SKF-89976-A Into Cultured Neurons and Atrocytes**

Substrate	K_m (µM)		Relative V_{max} (%)	
	Neurons	Astrocytes	Neurons	Astrocytes
GABA	25	30	100	100
Nipecotic acid	22	29	40	95
cis-4-OH-nip	46	25	35	50
ACHC	40	211	25	150
THPO	—	∞	—	—
SKF-89976 A	∞	∞	—	—

Mouse cerebral cortex neurons and astrocytes were cultured as detailed by Hertz et al. (1982; 1989a,b) and kinetic parameters for uptake of GABA and its analogues determined as described in the respective original papers. V_{max} values for GABA uptake into neurons and astrocytes are 2.8 and 0.9 nmol x min^{-1} x mg^{-1}, respectively. Values are from Larsson et al. (1980; 1983a,b; 1985; 1988).

nipecotic acid (SKF-89976-A) into cultured GABAergic neurons and astrocytes have revealed that distinct differences exist with regard to the ability of the GABA carriers to transport these compounds.

In neurons, nipecotic acid, *cis*-4-OH-nipecotic acid and ACHC are all substrates with K_m values similar to or slightly higher than that of GABA (Table I). This is compatible with the finding that these GABA analogues are

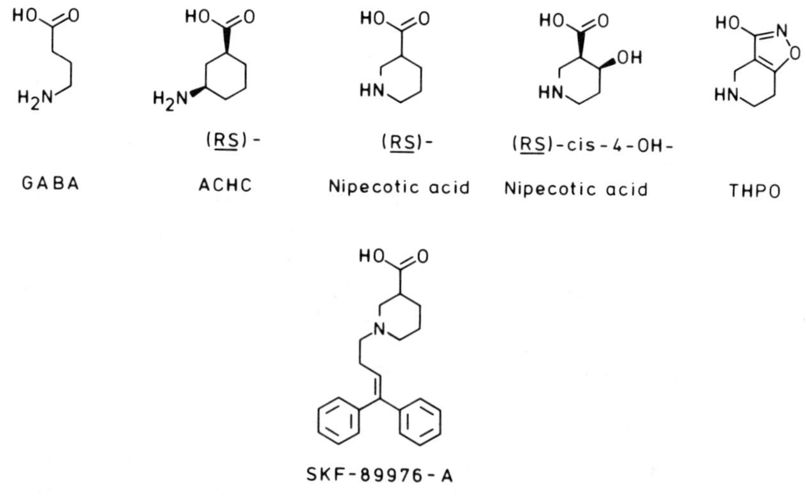

Fig. 1. Structural formulas of some key GABA analogues, which in radiolabeled forms have been used as tools for studies of GABA uptake mechanisms.

competitive inhibitors of GABA uptake (cf. below). It is, however, noteworthy that all 3 compounds were transported less efficiently than GABA, i.e., with much lower V_{max} values (Table I). It is particularly surprising that ACHC which was originally introduced as a selective neuronal GABA transport inhibitor (Bowery et al., 1976; Neal and Bowery, 1977; Neal et al., 1979) was found to be transported much less efficiently than GABA in the neurons. Judged from the similarity of K_m and K_i values for the mutual inhibition of GABA and ACHC (Table II) there is little doubt that the two compounds share the same binding site on the neuronal GABA carrier. The inability of the carrier to transport a competitive inhibitor is even more pronounced in case of SKF-89976-A since this compound was found not to be transported at all in a saturable fashion by the GABA carrier (Larsson et al., 1988).

As seen from Table I and Table III the uptake and mutual inhibition pattern of GABA and GABA analogues in astrocytes is more complicated than that observed in neurons (Table II). The most remarkable findings are that GABA and *cis*-4-OH-nipecotic acid are mutual non-competitive inhibitors indicating that although they are both transported with almost identical K_m values they are unlikely to interact in the same way with the GABA carrier. This may be reflected by the finding (Larsson et al., 1985) that V_{max} for *cis*-4-OH-nipecotic acid is only 50% of V_{max} for GABA uptake (Table I). The other surprising finding is that GABA is a competitive inhibitor of ACHC uptake whereas ACHC is a non-competitive inhibitor of GABA uptake (Larsson et al., 1983a). In spite of this, the efficiency (V_{max}) of ACHC uptake is higher than that for GABA uptake in astrocytes (Table I). These findings may indicate that more than one transport system for GABA exists in astrocytes (cf. Schousboe et al., 1983b).

In a discussion of the substrate requirements for neuronal and glial GABA carriers it seems essential to mention β-alanine which for almost 15

TABLE II. Mutual Inhibition by GABA and GABA Analogues of Uptake of the Same Amino Acids Into Cultured GABAergic Neurons

Amino acid as substrate	Amino acid as inhibitor			
	GABA	Nipecotic acid	*Cis*-4-OH-Nip	ACHC
GABA	25[a]	11	55	70
Nipecotic acid	15	22[a]	—	—
Cis-4-OH-Nip	24	—	46[a]	—
ACHC	20	—	—	40[a]

Mouse cerebral cortex neurons were cultured as detailed by Hertz et al. (1989b), and uptake and inhibition studies performed as described by Larsson et al. (1983a; 1985) which contain the original data. [a]Values represent K_i (μM) for competitive inhibition or K_m (μM) for uptake.

TABLE III. Mutual Inhibition by GABA and GABA Analogues of Uptake of the Same Amino Acids Into Cultured Astrocytes

Amino acid as substrate	Amino acid as inhibitor			
	GABA	Nipecotic acid	*Cis*-4-OH-Nip	ACHC
GABA	30[a]	15	147[*]	700[**]
Nipecotic acid	77	40[a]	—	—
Cis-4-OH-Nip	35[*]	—	25[a]	—
ACHC	15	—	—	211[a]

Mouse cerebral cortex astrocytes were cultured as detailed by Hertz et al. (1982; 1989a) and uptake and inhibition studies performed as described by Larsson et al. (1980; 1983a,b; 1985) which contain the original data. Values represent K_m (μM) for uptake [a] or K_i (μM) for competitive inhibition. One asterisk refers to a non-competitive inhibition type and two asterisks indicate an IC_{50} value (μM) since the inhibition kinetics were of a complex type.

years has been thought to be a selective inhibitor of glial GABA uptake (Schon and Kelly, 1975; Iversen and Kelly, 1975). In a recent investigation of the mutual inhibition pattern of β-alanine, GABA and taurine in neurons and astrocytes it was shown that GABA and β-alanine are mutual non-competitive inhibitors (Larsson et al., 1986b) in both neurons and astrocytes. On the other hand, β-alanine and taurine could be shown to share the same uptake site in both cell types (Table IV). The conclusion is that β-alanine is

TABLE IV. Mutual Inhibition by GABA, Taurine, and β-Alanine of the High-Affinity Uptake of the Same Amino Acids Into Cultured Neurons and Astrocytes

Amino acid as substrate	Amino acid as inhibitor					
	Neurons			Astrocytes		
	GABA	Taurine	β-Alanine	GABA	Taurine	β-Alanine
GABA	25[a]	15,500(N)	1,666(N)	31[a]	1,000[c]	843(N)
Taurine	911(N)[b]	20[a]	72(C)	>1,000[c]	25[a]	71(C)
β-Alanine	397(N)	217(C)	73[a]	1,222(N)	24(C)	71[a]

Mouse cerebral cortex neurons and astrocytes were cultured as detailed by Hertz et al. (1982; 1989a,b) and uptake experiments performed as described by Larsson et al. (1986a) which also contain the original data.
[a]K_m(μM) values for uptake of the respective amino acid.
[b]The values indicated with N or C represent K_i (μM) values for the inhibition; N, noncompetitive inhibition by the amino acid acting as an inhibitor; C, competitive inhibition by the amino acid acting as an inhibitor.
[c]Calculated from Schousboe et al. (1976) and Schousboe (1978). The type of inhibition is not clarified.

taken up into both neurons and astroglial cells and it uses in both cases the taurine carrier and not the GABA carrier. It is accordingly not advisable to use β-alanine to distinguish between neuronal and glial GABA uptake.

INHIBITION OF HIGH AFFINITY GABA UPTAKE
Structure-Activity Relations

Figure 2 illustrates the conformational flexibility of a number of GABA analogues with increasing rigidity of the molecule ranging from GABA with almost unlimited conformational flexibility to THPO and 5,6,7,8-tetrahydro-4H-isoxazolo[4,5-c]azepin-3-ol (THAO) which are almost totally rigid. The ability of these GABA analogues to inhibit neuronal and glial high affinity GABA uptake is summarized in Table V. It is seen that among the chiral molecules tested it is always the R-enantiomer which is able to inhibit GABA uptake. It is also noteworthy that the inhibitory potency is not dramatically influenced by the flexibility of the GABA analogues. Only in the extreme case of an inflexible molecule (THPO and THAO) is the inhibitory potency reduced. This reduction of the inhibitory potency is apparently more pronounced in the case of the neuronal GABA carrier than in the case of the astroglial carrier (cf. below). What seems to be much more important in terms of inhibitory potency is the relative distance between

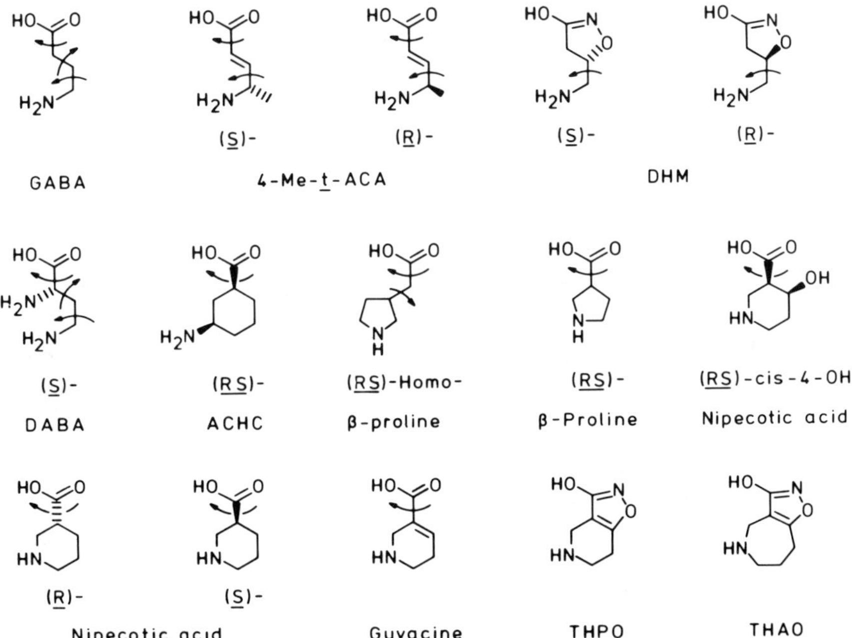

Fig. 2. Structural formulas with indications of structural conformational flexibility of GABA analogues which inhibit high affinity GABA uptake.

TABLE V. Inhibition of GABA Uptake by GABA Analogues With Decreasing Structural Conformational Flexibility

GABA analogue	Inhibition of GABA uptake IC$_{50}$ (µM)	
	Neuronal	Glial
GABA	15	35
(S)-DABA	1000	>5000
(RS)-ACHC	200	700
(R)-4-Me-GABA	200	120
(S)-4-Me-GABA	750	1000
(R)-4-Me-TACA	160	500
(S)-4-Me-TACA	>5000	>5000
(R)-DHM	800	2000
(S)-DHM	>5000	>5000
(RS)-Homo-β-proline	75	20
(RS)-β-proline	1200	320
(R)-Nipecotic acid	70	30
(S)-Nipecotic acid	500	1000
(RS)-Cis-4-OH-Nipecotic acid	200	10
Guvacine	200	25
THPO	5000	300
THAO	>5000	500

GABA uptake was studied at a ^{3}H-GABA concentration of 1 µM using either brain cortex mini-slices (Schousboe et al., 1979) or cultured neurons (Hertz et al., 1989b) to investigate neuronal GABA uptake, or cultured astrocytes (Hertz et al., 1982; 1989a) to investigate glial GABA uptake. Results are from studies of Schousboe et al. (1978; 1979; 1980; 1981) and Larsson et al. (1981).
Abbreviations: Me= Methyl; TACA= Transaminocrotonic acid; DHM= Dihydromuscimol.
For structural formulas of the GABA analogues, see Figure 2.

the charged centers of the molecules. This is illustrated in Figure 3 which shows the molecular structure of nipecotic acid, guvacine, THPO, isonipecotic acid, isoguvacine and 4,5,6,7-tetrahydroisoxazolo[5,4-c] pyridin-3-ol (THIP). The analogues with the shortest distance between the charged centers (nipecotic acid, guvacine, THPO) are potent inhibitors of GABA uptake (Table V) whereas those with a longer distance between the charged centers (isonipecotic acid, isoguvacine, THIP) are not (Schousboe et al., 1979). It should be stressed that these latter GABA analogues are potent GABA-A agonists and inhibitors of GABA-A receptor binding, whereas the former analogues have no effect on GABA receptor (Krogsgaard-Larsen, 1978). Moreover, the GABA receptor agonists THIP and isoguvacine are also not taken up into neurons or astrocytes by saturable transport processes (Schousboe et al., 1985).

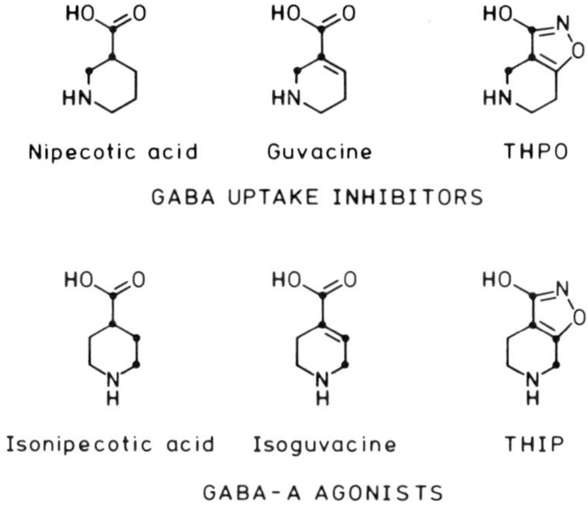

Fig. 3. Structural formulas of key GABA uptake inhibitors and GABA-A receptor agonists which differ with regard to the intramolecular distance between the functional groups.

Cellular Aspects

As pointed out above and previously by Schousboe et al.(1983b) the neuronal GABA carrier apparently exhibits a more strict structural specificity than the glial GABA carrier. This is indicated by the observation that when the flexibility of the GABA analogues becomes almost totally restrained as in the isoxazoles THPO and THAO (Fig. 2) the ability to inhibit GABA uptake is almost lost in the case of the neuronal carrier (Table V). In the case of the astrocytic carrier some of the inhibitory potency is lost but the compounds are still able to inhibit uptake at concentrations well below the millimolar range (Table V). This observation may be important in future attempts to design GABA analogues with which it is possible to selectively inhibit the glial GABA carrier.

In order to circumvent the problem of blood brain barrier penetration for the GABA uptake inhibitors nipecotic acid and guvacine, Yunger et al. (1984) and Ali et al. (1985) introduced the lipophilic substituent 4,4-diphenyl-3-butenyl (DPB) at the nitrogen atom of the parent amino acid. As can be seen from Table VI this chemical modification has increased the inhibitory potency of the amino acids considerably which is also in line with the original findings of Yunger et al. (1984). In light of the fact that these DPB derivatives are competitive inhibitors and therefore are likely to be recognized by the GABA carrier via the amino acid moiety (Ali et al., 1985; Larsson et al., 1988) it was somewhat surprising that DPB-nipecotic acid is not transported into neurons or astrocytes in a saturable manner *via* the GABA carrier. However, possibly owing to the lipophilic character of the molecule, DPB-nipecotic acid was found to enter the cells by a highly efficient, non-saturable process (Larsson et al., 1988). Recently, the inhibi-

**TABLE VI. Inhibition of GABA Uptake by GABA
Analogues of Restricted Conformation and Their DPB-Derivatives**

| | \multicolumn{8}{c}{K_i value (µM)} |
| | Neurons | | | | Astrocytes | | | |
	Nip.	Guv.	THPO	THAO	Nip.	Guv.	THPO	THAO
Parent amino acid	11	31	n.d.	n.d.	15	28	550	600
DPB-Derivative	1	5	38	9	2	4	26	3

Mouse cerebral cortex neurons and astrocytes were cultured as detailed by Hertz et al. (1982; 1989a,b) and K_i values (µM) for inhibition of ^{3}H-GABA uptake determined as described by Larsson et al. (1980; 1985; 1988) and Schousboe et al. (1981) which contain the original data.

n.d.= not determined since the IC_{50} values determined in brain cortex mini slices were 5000 µM or more (Schousboe et al., 1979).

Nip.: Nipecotic acid; Guv.: Guvacine.

tory potential of DPB derivatives of the glial selective GABA uptake inhibitors THPO and THAO has also been investigated (Krogsgaard-Larsen et al., 1987; Schousboe et al., 1990; Falch et al., 1990). As seen from Table VI these DPB derivatives were, like the nipecotic acid and guvacine analogues, found to be more potent inhibitors of GABA uptake than the parent compounds. This is particularly pronounced for DPB-THAO. However, the ability to distinguish between neuronal and glial GABA uptake has been lost by the incorporation of the lipophilic constituent in the molecules.

ANTICONVULSANT ACTIVITY OF GABA UPTAKE INHIBITORS
Penetration Through the Blood-Brain Barrier and the Use of Prodrugs

GABA and GABA analogues have a zwitterionic structure and the ratio between the respective pK_a values will indirectly determine the ability of the compound to penetrate the blood-brain-barrier. Since these compounds are most likely to penetrate the barrier in the unionized form, those with large differences in the two pK_a values are less likely to penetrate (Krogsgaard-Larsen et al., 1983). This is the case for GABA analogues such as nipecotic acid, cis-4-OH-nipecotic acid and guvacine. On the other hand, THPO with pK_a values of 4.3 and 9.1 is more likely to enter the brain after systemic administration, but even in this case only a very small fraction of THPO after intraperitoneal or intramuscular injection reaches the brain (Schousboe et al., 1986).

In order to circumvent this problem, prodrugs which reduce the hydrophilic character of the compounds have been prepared. The preferred method of prodrug preparation has been to synthesize esters such as ethyl-, octyl-, p-nitrophenyl-, and pivaloyloxymethyl-esters of nipecotic acid, cis-4-OH-nipecotic acid or guvacine (Krogsgaard-Larsen et al., 1987). That this method of increasing the permeability of the GABA analogues through the blood-brain barrier has been successful is shown by the repeated finding that the esters of the GABA analogues protect against different types of chemically or sound induced seizures after systemic administration (Horton et al., 1979; Frey et al., 1979; Krogsgaard-Larsen et al., 1981; Meldrum et al., 1982; Wood et al., 1983; Croucher et al., 1983). This is in contrast to the non-esterified amino acids which have anticonvulsant activity only after intracerebroventricular administration (Horton et al., 1979; Meldrum et al., 1982; Croucher et al., 1983). It is in keeping with this that intramuscular injection of the ethyl ester of nipecotic acid but not nipecotic acid itself leads to a significant increase in the synaptosomal neurotransmitter pool of GABA (Wood et al., 1980b), an effect closely related to the anticonvulsant action of either GABA uptake inhibitors (Schousboe et al., 1983a; 1986; Wood et al., 1983) or GABA transaminase inhibitors (Wood et al., 1980a; Gale et al., 1982; Gram et al., 1988). It should be mentioned that a drawback in the use of these prodrugs is the finding that the ethyl esters of guvacine and nipecotic acid have cholinergic properties (Frey et al., 1979; Zorn et al., 1987).

As mentioned above the preparation of N-substituted DPB derivatives of these amino acids constitutes an alternative way to circumvent the problem of blood-brain barrier penetration. It should be stressed, however, that this method does not represent a prodrug approach since the DPB derivatives are stable within the brain and the pharmacological action is brought about by the DPB derivative per se and not by the parent amino acid. This approach has proven highly effective in experimental animals since the DPB derivatives of nipecotic acid and guvacine are potent anticonvulsants (Yunger et al., 1984; Löscher, 1985; 1986; Zorn et al., 1986) which is in agreement with the potent inhibitory effect of these compounds on high affinity GABA uptake (Yunger et al., 1984; Ali et al., 1985; Larsson et al., 1988).

Relationship Between Anticonvulsant Activity and Inhibition of GABA Uptake at the Cellular Level

As pointed out above GABA released from nerve endings during neuronal activity is mainly taken up into presynaptic nerve endings (Hertz and Schousboe, 1987) where it can be reutilized as a neurotransmitter (Gram et al., 1988). Part of the released GABA is, however, taken up into surrounding astrocytes (Schousboe et al., 1977; Hertz and Schousboe, 1987) and accordingly lost from the neurotransmitter pool. In accordance with the view that reuptake of GABA into the presynaptic nerve ending is functionally important for the maintenance of the neurotransmitter related GABA pool, it has repeatedly been demonstrated that compounds like 2,4-diaminobutyric acid (DABA) and ACHC which preferentially inhibit the neuronal GABA carrier (Iversen and Kelly, 1975; Bowery et al., 1976) act as proconvulsants after intracerebroventricular administration (Horton et al., 1979; Meldrum et al., 1982; Gonsalves et al., 1989a). This may be explained by assuming that by preventing released GABA from gaining access to the neurotransmitter pool the inhibitory potential of GABA is decreased (Gale et al., 1982). The situation appears, however, somewhat more complex when the anticonvulsant properties of the group of GABA analogues inhibiting both the neuronal and the glial GABA uptake are considered. These are compounds like nipecotic acid and *cis*-4-OH-nipecotic acid where selectivity with regard to the uptake systems appears to be less pronounced. In some experimental seizure models these compounds protect against convulsions whereas in other models no effects are seen either after intracerebroventricular administration or administration as a prodrug via the systemic route (Horton et al., 1979; Frey et al., 1979; Meldrum et al., 1982; Croucher et al., 1983; Gonsalves et al., 1989a).

Based on the above mentioned considerations regarding the functional role of the neuronal and glial component of the mechanism responsible for the clearing of GABA from the synaptic cleft, it has been suggested (Schousboe, 1979; Krogsgaard-Larsen et al., 1981; Schousboe et al., 1983a,b) that the efficacy of GABA mediated neurotransmission might be enhanced if GABA could be prevented from being drained from the neu-rotransmission pool via uptake into the astroglial compartment. In other words, a selective blockade of the glial GABA carrier probably would have such an effect. With the use of a variety of experimental seizure models it has been shown (Krogsgaard-Larsen et al., 1981; Croucher et al., 1983; Wood et al., 1983; Schousboe et al., 1986; Gonsalves et al., 1989a,b) that the glial selective GABA uptake inhibitors THPO and THAO (Schousboe et al., 1981) act as anticonvulsants provided they get access to the brain. The relationship between the actual brain concentration of THPO and the ability of THPO to protect against seizures induced by isonicotinic acid hydrazide (INH) inhibiting GABA synthesis has been investigated (Schousboe et al., 1986). As seen from Table VII a brain level of THPO of about 500 μM is required to protect against such seizures. This is compatible with the previous finding that the K_i value for THPO inhibition of GABA uptake into astrocytes is almost identical to this tissue concentration (Table VI, VII).

Recently, Gonsalves et al. (1989a,b) using identical conditions tested the anticonvulsant activity of different GABA transport inhibitors belonging to the above mentioned categories of neuron-selective, mixed or glia-selective inhibitors. As seen from Table VIII both THPO and THAO protected against INH induced seizures whereas *cis*-4-OH-nipecotic acid and DABA had no effect. In the pentylenetetrazol (PTZ) seizure model, however, both THPO, THAO and DABA protected whereas *cis*-4-OH-nipecotic acid had no effect, as in the former case. To facilitate a comparison of the pharmacological action of these GABA uptake inhibitors a summary of the results is provided in Table IX. It is seen that only the glial selective GABA uptake inhibitors THPO and THAO are effective anticonvulsant agents in both seizure models. Moreover, none of these compounds exhibited proconvulsant activity whereas the neuronal uptake inhibitor DABA exhibited such activity in agreement with previous observations (Horton et al., 1979; Meldrum et al., 1982). Gonsalves et al., (1989b) also compared some of the side effects seen after intracerebroventricular administration of THPO and THAO. As seen

TABLE VII. Comparison of Brain Levels and Anticonvulsant Action of THPO After Systemic Administration

Time after injection (hr)	THPO in brain (μmol x g^{-1})		Time to onset of INH-seizures (min)	
	Mice	Chicks	Mice	Chicks
0	—	—	23.8	27.6±1.3
1	0.116	0.512	25.0	40.8±1.2*
3	0.088	0.515	25.8	38.0±1.7*

Animals were injected intramuscularly with THPO or [^{3}H]THPO (1 mmol x kg^{-1}) and seizures were induced by injection of 2.2 mmol x kg^{-1} INH. Number of animals in each group was 16 and results are averages ± SEM. *Statistically significant differences from the control group (P <0.001). (Reproduced from Schousboe et al. (1986), with permission of publisher.)

TABLE VIII. Effects of GABA Mimetics on INH-Induced Generalized Motor Seizures and PTZ-Induced Seizures

Treatment	Dose (μg, i. c. v.)	INH seizure latency	Maximal PTZ seizures	
			% protected	% survived
CSF		36.2±2.5	5	20
THPO	100	48.7±6.0*	0	33
	300	46.8±3.5*	88*	100*
THAO	100	42.7±4.0	—	—
	200	51.8±4.6*	0	38
	300	51.8±3.9*	38	38
	750	45.8±3.9	77*	92*
Cis-4-OH-ni-	30	31.6±5.7	17	33
pecotic acid	300	33.2±4.5	33*	50
DABA	1500	35.4±3.7	92**	92**

Drugs or mock CSF vehicle were infused i.c.v. 30 min before INH (822 mg/kg, i.v.) or i.v. bolus injection of PTZ (25 mg/kg). The latency (min ± S.E.) to convulsion is indicated for the number of animals tested. Significance of differences for INH seizures compared with CSF-treated group (control values pooled from 3 independent groups) is indicated by asterisks (*P <0.05, **P <0.001). Percentages (PTZ seizures) reflect protection against tonic forelimb extension and death. Control groups were pooled for statistical analysis. Asterisks in this group indicate significance of differences compared with CSF control by chi-square statistic (*P <0.05, **P <0.01). (Reproduced from Gonsalves et al. (1989a,b), with permission of publisher.)

TABLE IX. Semiquantitative Summary of Antiseizure Effects of GABAmimetics

Treatment	Seizure model		Proconvulsant activity
	Max PTZ	INH	
THPO	++++	+	No
THAO	++++	+	No
Cis-4-OH-ni-pecotic acid	0	0	No
DABA	++++	0	Yes

The antiseizure effects detailed in Table VIII are summarized. For maximal PTZ seizures, only compounds protecting at least 50% of animals against the tonic extensor component were considered anticonvulsant. Degree of anticonvulsant activity is indicated as follows: + = ≤25% increase over CSF control; ++ = >25%, ≤50% increase; +++ = >50%, ≤100% increase; ++++ = >100% increase. Proconvulsant activity was recorded as present or absent. (Reproduced from Gonsalves et al. (1989a,b), with permission of publisher.)

in Table X, THAO exhibited significantly fewer adverse effects than THPO. THAO may accordingly serve as a lead structure in future attempts to develop anticonvulsant GABA analogues whose action is based on a preferential inhibition of glial GABA uptake.

Table X. Incidence of Neurotoxicity Following i.c.v. Injection of THAO (750 μg) or THPO (300 μg) in Rats

Behavior observed	Incidence of response (%)	
	THAO	THPO
Reduced/absent locomotor activity	0[a]	89
Abnormal gait or stance	11	56
Body sag	0[a]	67
Decreased muscle tone	44[a]	89
Spontaneous twitches	22	22
Exaggerated startle reflex	44	11

The doses selected had been shown previously to protect at least 75% of rats against maximal PTZ seizures. The data are expressed as the percentage of animals (N = 9 rats/group) exhibiting a given behavior. [a]$P<0.05$ by the Fisher exact test compared with THPO-treated group. (Reproduced from Gonsalves et al. (1989b), with permission of publisher.)

Synergistic Actions of Glycine

It was originally reported by Seiler and Sarhan (1984a,b) and Sarhan et al., (1984) that systematically co-administered glycine enhanced the anticonvulsant effect of either GABA-transaminase inhibitors such as gamma-vinyl GABA or GABA receptor agonists such as muscimol. This work was subsequently extended to include the glial GABA uptake inhibitor THPO (Seiler et al., 1985). As seen from Table XI co-administration of glycine (10 mmol/kg subcutaneously) greatly enhanced the ability of intraperitoneally administered THPO to protect rats against seizures induced by 3-mercaptopropionic acid, an inhibitor of glutamate decarboxylase (Lamar, 1970; Wu and Roberts, 1974). It was also observed that glycine had a similar synergistic action in PTZ induced seizures (Seiler et al., 1985). Administered alone, glycine had no effect in any of the seizure models. One simple way of explaining the action of glycine would be to postulate that it enhances the uptake of THPO into the brain. This was, however, shown by the same authors not to be the case. In an attempt to clarify the mechanism by which glycine enhances the anticonvulsant efficacy of THPO, Wood et al. (1988) studied the effects of glycine and THPO on synaptosomal GABA levels. As seen in Figure 4 co-administration of glycine and THPO led to synaptosomal GABA levels which

**TABLE XI. Anticonvulsant Effect of THPO
and of Combined Treatment With THPO and Glycine
on 3-Mercaptopropionic Acid (MPA)-Induced Convulsions in Mice**

Treatment					
THPO (mmol/kg i.p.)	Glycine (mmol/kg s.c.)	N^1	N^2	N^3	% of animals with seizures
0	0	10	5	14	100
2	0	9	6	15	90
4	0	9	5	10	90*
0	10	9	0	12	90
2	10	6	1	6	60
4	10	4	0	4	40*

Each treatment group consisted of 10 mice. THPO was given first (intraperitoneally). Fifteen minutes later the animals received either physiological saline or glycine (subcutaneously). Seizures were induced 45 min after glycine administration (60 min after THPO) by i.p. injection of 40 mg/kg MPA.
N^1 = Number of animals with clonic convulsions.
N^2 = Number of animals with tonic hind limb extension.
N^3 = Number of clonic seizure episodes.
The asterisks indicate a statistically significant ($P = 0.05$) difference between the two groups, one treated with THPO alone, the other treated with THPO and glycine (glycine effect). (Non-parametric statistics.) (Reproduced from Seiler et al. (1985), with permission of publisher.)

are higher than those reached by administration of either THPO or glycine alone. The highest synaptosomal GABA levels are, however, still lower than those normally required to give protection against isonicotinic acid hydrazide induced convulsions (Wood et al., 1980b). It is accordingly unlikely that the effect of glycine can be explained by a direct action on the availability of GABA in the neurotransmitter pool. It should, however, be noted that co-administration of glycine and THPO also led to a significant reduction of the synaptosomal glutamate content (Wood et al., 1988) which conceivably could affect excitatory neurotransmission processes mediated by glutamate. This could in turn lead to an altered balance between the excitatory and inhibitory neurotransmission processes. An alternative explanation of this peculiar action of glycine in conjunction with THPO may be offered by the demonstration that THPO has a weak antagonistic action on glycine receptors (Krogsgaard-Larsen et al., 1975; 1983). Administration of glycine would prevent such actions of THPO which, in turn, would make the enhancing effect of THPO on the GABA mediated neurotransmission more prominent (Krogsgaard-Larsen et al.,

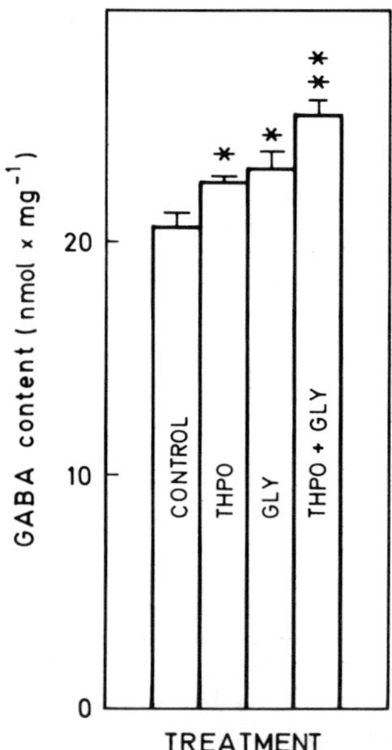

Fig. 4. Synaptosomal content of GABA (nmol x mg^{-1}) after treatment of mice with THPO and/or glycine. Treatment consisted of intramuscular injection of THPO (2 mmol/kg) and/or glycine (10 mmol/kg). Synaptosomes were prepared 60 min after injection of THPO. Asterisks indicate statistically significant differences from the control (*$P < 0.05$; **$P < 0.001$). All groups consisted of 6 animals except the THPO treated which consisted of 4 animals. (Reproduced from Wood et al. (1988), with permission of publisher.)

1987). Whatever the exact mechanism may be, it may have interesting pharmacological and therapeutic aspects that glycine synergistically enhances the anticonvulsant properties of GABAergic drugs.

CONCLUSIONS

As discussed in detail in the different chapters of this book many different approaches have been taken to manipulate the GABA mediated neurotransmission processes with the aim to achieve an antiseizure effect. As shown in this review one such approach is inhibition of high affinity GABA uptake which may result in a facilitation of the action of GABA at its receptors. The demonstration that inhibitors of high affinity GABA transport which preferentially act on the transport system present in astroglial cells are possibly more efficient anticonvulsant agents than inhibitors acting preferentially on neuronal uptake sites may have some interesting implications. It may be concluded that the glial component of high affinity GABA transport may be an important factor in the normal, physiological control of the inactivation of GABA at the synaptic level. Although, based on circumstantial evidence, this has for some time been assumed to be the case, it has never been possible to get a direct proof for this functional property of the astrocytes. From a pharmacological point of view this observation may also be of interest. Using simple cell culture models it may be possible to predict whether a given GABA analogue may have anticonvulsant properties. It is likely to be less tedious, costly and time consuming to use inhibition of GABA uptake into cultured astrocytes as a preliminary screening system than to use a variety of in vivo seizure models.

ACKNOWLEDGMENTS

The experimental work has been supported by grants from the Danish Biotechnology Program (1987–1990) and the Lundbeck Foundation. The secretarial assistance of Mrs. Hanne Danø is highly appreciated.

REFERENCES

Ali FE, Bondinell WE, Dandridge PA, Frazee JS, Garvey E, Girard GR, Kaiser C, Ku TW, Lafferty JJ, Moonsammy GI, Oh H-J, Rush JA, Setler PE, Stringer OD, Venslavsky JW, Volpe BW, Yunger LM, Zircle CL: Orally active and potent inhibitors of γ-aminobutyric acid uptake. J Med Chem 28:653–660, 1985.

Bowery NG, Jones GP, Neal MJ: Selective inhibition of neuronal GABA uptake by cis-1,3-aminocyclohexane carboxylic acid. Nature 264:281–284, 1976.

Croucher MJ, Meldrum BS, Krogsgaard-Larsen P: Anticonvulsant activity of GABA uptake inhibitors and their prodrugs following central or systemic administration. Eur J Pharmacol 89:217–228, 1983.

Curtis DR and Johnston GAR: Amino acid transmitters. In Lajtha A (ed) "Handbook of Neurochemistry," 1st ed., Vol. 4. New York: Plenum Press, 1970, pp 115–134.

Curtis DR, Johnston GAR: Amino acid transmitters in the mammalian central nervous system. Ergeb Physiol 69:97–188, 1974.

Curtis DR, Grame CJA, Lodge D: The in vivo inactivation of GABA and other inhibitory amino acids in the cat nervous system. Exp Brain Res 25:413–428, 1976.

Drejer J, Meier E, Schousboe A: Novel neuron-related regulatory mechanisms for astrocytic glutamate and GABA high affinity uptake. Neurosci Lett 37:301–306, 1983.

Elliott KAC, van Gelder NM: Occlusion and metabolism of gamma-aminobutyric acid by brain tissue. J Neurochem 3:28–40, 1958.

Fagg GE, Foster AC: Amino acid neurotransmitters and their pathways in the mammalian central nervous system. Neuroscience 9:701–719, 1983.

Falch E, Larsson OM, Schousboe A, Krogsgaard-Larsen P: GABA-A agonists and GABA uptake inhibitors: Structure-activity relationships. Drug Dev Res 21: 169–188, 1990.

Fariello RG, Morselli PL, Lloyd KG, Quesney LF, Engel J: "Neurotransmitters, Seizures, and Epilepsy," Vol. II. New York: Raven Press, 1984, 371 pp.

Frey H-H, Popp C and Löscher W: Influence of inhibitors of the high affinity GABA uptake on seizure thresholds in mice. Neuropharmacology 18:581–590, 1979.

Gale K, Iadarola MJ, Casu M, Keating RF: Relationship between GABA levels in vivo and anticonvulsant activity: importance of cellular compartments and regional localization in brain. In Okada Y and Roberts E (eds): "Problems in GABA Research from Brain to Bacteria." Amsterdam: Excerpta Medica, 1982, pp 159–172.

Gonsalves S, Twitchell B, Harbaugh RE, Krogsgaard-Larsen P, Schousboe A: Anticonvulsant activity of intracerebroventricularly administered glial GABA uptake inhibitors and other GABAmimetics in chemical seizure models. Epilepsy Res 4:34–41, 1989a.

Gonsalves S, Twitchell B, Harbaugh RE, Krogsgaard-Larsen P, Schousboe A: Anticonvulsant activity of the glial GABA uptake inhibitor, THAO, in chemical seizures. Eur J Pharmacol 168:265–268, 1989b.

Gram L: Experimental studies and controlled clinical testing of valproate and vigabatrin. Acta Neurol Scand 78:241–270, 1988.

Gram L, Larsson OM, Johnsen AH, Schousboe A: Effects of valproate, vigabatrin and aminooxyacetic acid on release of endogenous and exogenous GABA from cultured neurons. Epilepsy Res 2:87–95, 1988.

Hertz E, Yu ACH, Hertz L, Juurlink BHJ, Schousboe A: Preparation of primary cultures of mouse cortical neurons. In Shahar A, DeVellis J, Vernadakis A and Haber B (Eds): "A Dissection and Tissue Culture Manual of the Nervous System." New York: Alan R Liss, Inc., 1989a, pp 183–186.

Hertz L: Functional interactions between neurons and astrocytes I. Turnover and metabolism of putative amino acid transmitters. Prog Neurobiol 13:277–323, 1979.

Hertz L, Schousboe A: Role of astrocytes in compartmentation of amino acid and energy metabolism. In Fedoroff S and Vernadakis A (eds): "Astrocytes," Vol. 2. New York: Academic Press, Inc., 1986, pp. 179–208.

Hertz L, Schousboe A: Primary cultures of GABAergic neurons as model systems to study neurotransmitter functions. I. Differentiated cells. In Vernadakis A, Privat A, Lauder JM, Timiras PS and Giacobini E (eds): "Model Systems of Development and Aging of the Nervous System." Boston: Martinus Nijhoff Publishing, 1987, pp 19–31.

Hertz L, Juurlink BHJ, Fosmark H, Schousboe A: Astrocytes in primary cultures. In Pfeiffer SE (ed): "Neuroscience Approached Through Cell Culture," Vol. 1. Florida: CRC-Press, Boca Raton, 1982, pp 175–186.

Hertz L, Drejer J and Schousboe A: Energy metabolism in glutamatergic neurons, GABAergic neurons and astrocytes in primary cultures. Neurochem Res 13:605–610, 1988.

Hertz L, Juurlink BHJ, Hertz E, Fosmark H, Schousboe A: Preparation of primary cultures of mouse (rat) astrocytes. In Shahar A, DeVellis J, Vernadakis A and Haber B (Eds): "A Dissection and Tissue Culture Manual of the Nervous System." New York: Alan R Liss, Inc., 1989b, pp 105–108.

Horton RW, Collins JF, Anlezark GM, Meldrum BS: Convulsant and anticonvulsant actions in DBA/2 mice of compounds blocking the reuptake of GABA. Eur J Pharmacol 59:75–83, 1979.

Hösli E, Schousboe A, Hösli L: Amino acid uptake. In Fedoroff S and Vernadakis A (eds): "Astrocytes," Vol 2. New York: Academic Press, Inc., 1986, pp 133–153.

Hydén H, Cupello A, Palm A: Assymetric diffusion into the postsynaptic neuron: an extremely efficient mechanism for removing excess GABA from synaptic clefts on the Deiter's neurone plasma membrane. Neurochem Res 11:695–706.

Iversen LL: The uptake, storage, release and metabolism of GABA in inhibitory nerve. In Snyder H (ed): "Perspectives in Neuropharmacology." London and New York: Oxford University Press, 1972, pp 75–111.

Kanner BI: Active transport of γ-aminobutyric acid by membrane vesicles isolated from rat brain. Biochemistry 17:1207–1211, 1978.

Kanner BI, Bendahan A, Radian R: Efflux and exchange of gamma-aminobutyric acid and nipecotic acid catalysed by synaptic plasma membrane vesicles isolated from immature rat brain. Biochim Biophys Acta 731:54–62, 1983.

Keynan S, Kanner BI: γ-Aminobutyric acid transport in reconstituted preparations from rat brain: Coupled sodium and chloride fluxes. Biochemistry 27:12–17, 1988.

Krogsgaard-Larsen P: GABA agonists and uptake inhibitors of restricted conformations: structure-activity relations. In Fonnum F (ed): "Amino Acids as Chemical Transmitters." New York: Plenum Press, 1978, pp 305–321.

Krogsgaard-Larsen P: γ-Aminobutyric acid agonists, antagonists, and uptake inhibitors. Design and therapeutic aspects. J Med Chem 24:1377–1383, 1981.

Krogsgaard-Larsen P, Johnston GAR, Curtis DR, Game CJA, McCulloch R: Structure and biological activity of a series of conformationally restricted analogues of GABA. J Neurochem 25:803–809, 1975.

Krogsgaard-Larsen P, Labouta I, Meldrum B, Croucher M, Schousboe A: GABA uptake inhibitors as experimental tools and potential drugs in epilepsy research. In Morselli PL, Reynolds EM, Lloyd KG, Lösher W and Meldrum BS (eds): "Neurotransmitters, Seizures and Epilepsy." New York: Raven, 1981, pp 23–33.

Krogsgaard-Larsen P, Mikkelsen H, Jacobsen P, Falch E, Curtis DR, Peet MJ, Leah JD: 4,5,6,7-Tetrahydroisothiazolo[5,4-c]pyridin-3-ol and related analogues of THIP. Synthesis and biological activity. J Med Chem 26:895–900, 1983.

Krogsgaard-Larsen P, Falch E, Larsson OM, Schousboe A: GABA uptake inhibitors: relevance to antiepileptic drug research. Epilepsy Res 1:77–93, 1987.

Krogsgaard-Larsen P, Hjeds H, Falch E, Jørgensen FS, Nielsen L: Recent advances in GABA agonists, antagonists and uptake inhibitors: Structure-activity relationships and therapeutic potential. Adv Drug Res 17:381–456, 1988.

Lamar C: Mercaptopropionic acid: A convulsant that inhibits glutamate decarboxylase. J Neurochem 17:165–170, 1970.

Larsson OM, Krogsgaard-Larsen P, Schousboe A: High-affinity uptake of (RS)-nipecotic acid in astrocytes cultured from mouse brain. Comparison with GABA transport. J Neurochem 34:970–977, 1980.

Larsson OM, Thorbek P, Krogsgaard-Larsen P, Schousboe A: Effect of homo-β-proline and other heterocyclic GABA analogues on GABA uptake in neurons and astroglial cells and on GABA receptor binding. J Neurochem 37:1509–1516, 1981.

Larsson OM, Johnston GAR, Schousboe A: Differences in uptake kinetics of *cis*-3-aminocyclohexane carboxylic acid into neurons and astrocytes in primary cultures. Brain Res 260:279–285, 1983a.

Larsson OM, Drejer J, Hertz L, Schousboe A: Ion dependency of uptake and release of GABA and (*RS*)-nipecotic acid studied in cultured mouse brain cortex neurons. J Neurosci Res 9:291–302, 1983b.

Larsson OM, Krogsgaard-Larsen P, Schousboe A: Characterization of uptake of GABA, nipecotic acid and *cis*-4-OH-nipecotic acid in cultured neurons and astrocytes. Neurochem Int 7:853–860, 1985.

Larsson OM, Hertz L, Schousboe A: Uptake of GABA and nipecotic acid in astrocytes and neurons in primary cultures: Changes in the sodium coupling ratio during differentiation. J Neurosci Res 16:699–708, 1986a.

Larsson OM, Griffiths R, Allen IC, Schousboe A: Mutual inhibition kinetic analysis of γ-aminobutyric acid, taurine, and β-alanine high-affinity transport into neurons and astrocytes: Evidence for similarity between the taurine and β-alanine carriers in both cell types. J Neurochem 47:426–432, 1986b.

Larsson OM, Falch E, Krogsgaard-Larsen P: Kinetic characterization of inhibition of γ-aminobutyric acid uptake into cultured neurons and astrocytes by 4,4-diphenyl-3-butenyl derivatives of nipecotic acid and guvacine. J Neurochem 50:818–823, 1988.

Lloyd KG, Munari C, Bossi L, Morselli PL: GABA hypothesis of human epilepsy: neurochemical evidence from surgically resected identified foci. In Fariello RG, Morselli PL, Lloyd KG, Quesney LF and Engel J (eds): "Neurotransmitters, Seizures, and Epilepsy," Vol. II. New York: Raven Press, 1984, pp 285–291.

Löscher W: Anticonvulsant action in the epileptic gerbil of novel inhibitors of GABA uptake. Eur J Pharmacol 110:103–108, 1985.

Löscher W: SK&F-89976-A. Drugs of the Future 11:36–38, 1986.

Löscher W, Siemes H: Cerebrospinal fluid gamma-amino-butyric acid in children with different types of epilepsy: effect of anticonvulsant treatment. Epilepsia 26:314–319, 1985.

Martin DL: Carrier-mediated transport and removal of GABA from synaptic regions. In Roberts E, Chase TN and Tower DB (eds): "GABA in Nervous System Function." New York: Raven Press, 1976, pp 347–386.

Meldrum BS: Epilepsy and γ-aminobutyric acid-mediated inhibition. Int Rev Neurobiol 17:1–36, 1975.

Meldrum BS, Croucher MJ, Krogsgaard-Larsen P: GABA-uptake inhibitor as anticonvulsant agents. In Okada Y and Roberts E (eds): "Problems in GABA Research from Brain to Bacteria." Amsterdam: Excerpta Med, 1982, pp 501–533.

Morselli PL, Lloyd KG, Löscher W, Meldrum B, Reynolds EH: "Neurotransmitters, Seizures and Epilepsy." New York: Raven Press, 1981, 361 pp.

Neal MJ, Bowery NG: *Cis*-3-aminocyclohexane carboxylic acid: a substrate for the neuronal GABA transport system. Brain Res 138:169–174, 1977.

Neal MJ, Cunningham JR and Marshall J: The uptake and radioautographical localization in the frog retina of [^{3}H]-(±)-aminocyclohexane carboxylic acid, a selective inhibitor of neuronal GABA transport. Brain Res 176:285–296, 1979.

Nistico G, Engel J, Fariello RG, Lloyd KG, Morselli PL: "Neurotransmitters, Seizures, and Epilepsy," Vol. III. New York: Raven Press, 1986, 505 pp.

Olsen RW, Venter JC: "Benzodiazephine/GABA Receptors and Chloride Channels, Structural and Functional Properties." New York: Alan R. Liss, 1986, 351 pp.

Ribak CE: Axon terminals of GABAergic chandelier cells are lost at epileptic foci. Brain Res 326:217–225, 1985.

Roberts E: New directions in GABA research. I. Immunocytochemical studies of GABA neurons. In Krogsgaard-Larsen P, Scheel-Krüger J and Kofod H (eds): GABA-neurotransmitters. Pharmacochemical, Biochemical and Pharmacological Aspects." Copenhagen: Munksgaard, 1979, pp 28–45.

Ross SM, Craig CR: Studies on γ-aminobutyric acid transport in cobalt experimental epilepsy in the rat. J Neurochem 36:1006–1011, 1981.

Sarhan S, Kolb M, Seiler N: The amplification of the anticonvulsant effect of vinyl GABA (4-aminohexenoic acid) by esters of glycine. Drug Res 34:687–690, 1984.

Schon F, Kelly JS: Selective uptake of β-[^{3}H]alanine by glia: Association with the glial uptake system for GABA. Brain Res 86:243–257, 1975.

Schousboe A: "Metabolism and function of GABA in the mammalian brain. Possible role of glia cells." D.Sc. Thesis. Copenhagen: University of Copenhagen, 1978, pp 1–83.

Schousboe A: Effects of GABA analogs on the high-affinity uptake of GABA in astrocytes in primary culture. In Mandel P and De Feudis FV (eds): "GABA-Biochemistry and CNS Function." New York: Plenum, 1979, pp 219–237.

Schousboe A: Transport and metabolism of glutamate and GABA in neurons and glial cells. Int Rev Neurobiol 22:1–45, 1981.

Schousboe A: Metabolism and function of neurotransmitters. In Pfeiffer SE (ed): "Neuroscience Approached through Cell Culture," Vol I. Florida: CRC Press, Boca Raton, 1982, pp 106–141.

Schousboe A: Neurochemical alterations associated with epilepsy or seizure activity. In Dam M and Gram L (eds): "Comprehensive Epileptology." New York: Raven Press, 1990, pp 1–16.

Schousboe A, Fosmark H and Svenneby G: Taurine uptake in astrocytes cultured from dissociated mouse brain hemispheres. Brain Res 116:158–164, 1976.

Schousboe A, Svenneby G and Hertz L: Uptake and metabolism of glutamate in astrocytes cultured from dissociated mouse brain hemispheres. J Neurochem 29:999–1005, 1977.

Schousboe A, Krogsgaard-Larsen P, Svenneby G, Hertz L: Inhibition of the high-affinity, net uptake of GABA into cultured astrocytes by β-proline, nipecotic acid and other compounds. Brain Res 153:623–626, 1978.

Schousboe A, Thorbek P, Hertz L and Krogsgaard-Larsen P: Effects of GABA analogues of restricted conformation on GABA transport in astrocytes and brain cortex slices and on GABA receptor binding. J Neurochem 33:181–189, 1979.

Schousboe A, Hertz L, Larsson OM, Krogsgaard-Larsen P: Interactions between conformationally restricted GABA analogues and the astroglial GABA carrier. Brain Res Bull 5 [Suppl 2]:403–409, 1980.

Schousboe A, Larsson OM, Hertz L, Krogsgaard-Larsen P: Heterocyclic GABA analogues as new selective inhibitors of astroglial GABA transport. Drug Devel Res 2:115–127, 1981.

Schousboe A, Larsson OM, Wood JD, Krogsgaard-Larsen P: Transport and metabolism of gamma-aminobutyric acid in neurons and glia: implications for epilepsy. Epilepsia 24: 531–538, 1983a.

Schousboe A, Larsson OM, Drejer J, Krogsgaard-Larsen P, Hertz L: Uptake and release processes for glutamine, glutamate and GABA in cultured neurons and astrocytes. In Hertz L, Kvamme E, McGeer EG and Schousboe A (eds): "Glutamine, Glutamate, and GABA in the Central Nervous System". New York: Alan R. Liss, Inc., 1983b, pp 297–315.

Schousboe A, Larsson OM and Krogsgaard-Larsen P: Lack of a high affinity uptake system for the GABA agonists THIP and isoguvacine in neurons and astrocytes cultured from mouse brain. Neurochem Int 7:505–508, 1985.

Schousboe A, Hjeds H, Engler J, Krogsgaard-Larsen P, Wood JD: Tissue distribution, metabolism, anticonvulsant efficacy, and effect on brain amino acid levels of the glia-selective γ-aminobutyric acid transport inhibitor 4,5,6,7-tetrahydroisoxazolo[4,5-c]pyridin-3-ol in mice and chicks. J Neurochem 47:758–763, 1986.

Schousboe A, Drejer J, Larsson OM, Meier E: Regulation of astrocytic high affinity uptake of transmitter amino acids by neuronal signaling. In Althaus HH and Seifert W (eds): "Glial-Neuronal Communication in Development and Regeneration," Vol 2. Berlin: Springer-Verlag, pp 233–245, 1987.

Schousboe A, Larsson OM, Krogsgaard-Larsen P, Drejer J, Hertz L: Uptake and release processes for neurotransmitter amino acids in astrocytes. In Norenberg MD, Hertz L and Schousboe A (eds): "The Biochemical Pathology of Astrocytes." New York: Alan R. Liss, Inc., 1988, pp 381–394.

Schousboe A, Krogsgaard-Larsen P, Larsson OM, Gonsalves SF, Harbaugh RE, Wood JD: GABA uptake inhibitors: Possible use as antiepileptic drugs. In Lubec G and Rosenthal GA (eds): "Amino Acids. Chemistry, Biology and Medicine," Leiden: ESCOM Publ., 1990, pp 345–351.

Seiler N, Sarhan S: Synergistic anticonvulsant effects of GABA-T inhibitors and glycine. Naunyn Schmiedebergs Arch Pharmacol 326:49–57, 1984a.

Seiler N, Sarhan S: Synergistic anticonvulsant effects of a GABA agonist and glycine. Gen Pharmacol 15:367–369, 1984b.

Seiler N, Sarhan S, Krogsgaard-Larsen P, Hjeds H, Schousboe A: Amplification by glycine of the anticonvulsant effect of THPO, a GABA uptake inhibitor. Gen Pharmacol 16:509–511, 1985.

Sellström Å, Venema R, Henn F: Functional assessment of GABA uptake or exchange by synaptosomal fractions. Nature 264:652–653, 1976.

Snyder SH, Kuhar MJ, Green AI, Coyle JT, Shaskan EG: Uptake and subcellular localization of neurotransmitters in the brain. Int Rev Neurobiol 13:127–158, 1970.

Spyrou NA, Prestwich SA, Horton RW: Synaptosomal [^{3}H]GABA uptake and [^{3}H]nipecotic acid binding in audiogenic seizure susceptible (DBA/2) and resistant (C57 BL/6) mice. Eur J Pharmacol 100:207–210, 1984.

Wood JD: The role of γ-aminobutyric acid in the mechanism of seizures. Prog Neurobiol 5:79–95, 1975.

Wood JD, Russell MP, Kurylo E: The γ-aminobutyrate content of nerve endings (synaptosomes) in mice after the intramuscular injection of γ-aminobutyrate-elevating agents: A possible role in anticonvulsant activity. J Neurochem 35:125–130, 1980a.

Wood JD, Schousboe A, Krogsgaard-Larsen P: In vivo changes in the GABA content of nerve endings (synaptosomes) induced by inhibitors of GABA uptake. Neuropharmacology 19:1149–1152, 1980b.

Wood JD, Johnson DD, Krogsgaard-Larsen P, Schousboe A: Anticonvulsant activity of the glial-selective GABA uptake inhibitor, THPO. Neuropharmacology 22:139–142, 1983.

Wood JD, Krogsgaard-Larsen P, Schousboe A: Amplification by glycine of the effect of the GABA transport inhibitor THPO on synaptosomal GABA level. Neurochem Res 13:917–921, 1988.

Wood JH, Hare TA, Glaeser BS, Ballenger JC and Post RM: Low cerebrospinal fluid γ-aminobutyric acid content in seizure patients. Neurology (Minneapolis) 29:1203–1208, 1979.

Wu J-Y and Roberts E: Properties of brain L-glutamate decarboxylase: Inhibition studies. J Neurochem 23:759–767, 1974.

Yu ACH and Hertz L: Metabolic sources of energy in astrocytes. In Hertz L, Kvamme E, McGeer EG and Schousboe A (eds): "Glutamine, Glutamate and GABA in the Central Nervous System." New York: Alan R. Liss, Inc, 1983, pp 431–438.

Yunger LM, Fowler PJ, Zarevics P, Setler PE: Novel inhibitors of gamma-aminobutyric acid (GABA) uptake: anticonvulsant actions in rats. J Pharmacol Exp Ther 228:109–115, 1984.

Zorn SH, Willmore LJ, Bailey CM, Enna SJ: A comparison of the antinociceptive and anticonvulsant effects of GABAergic drugs: Evidence for a GABA receptor system unrelated to GABA$_A$ or GABA sites. In Nistico G, Morselli PL, Lloyd KG, Fariello RG and Engel J (eds): "Neurotransmitters, Seizures, and Epilepsy," Vol. III. New York: Raven Press, 1986, pp 123–133.

Zorn SH, Duman RS, Giachetti A, Micheletti R, Giraldo E, Krogsgaard-Larsen P, Enna SJ: (R)-Nipecotic acid ethyl ester: a direct-acting cholinergic agonist that appears selective for a subclass of muscarinic receptors. J Pharmacol Exp Ther 242:173–178, 1987.

GABA Mechanisms in Epilepsy, pages 189–204
© 1991 Wiley-Liss, Inc.

9
GABA Aminotransferase Inhibitors as Potential Antiepileptic Agents

GODFREY TUNNICLIFF

Laboratory of Neurochemistry, Indiana University School of Medicine, Evansville, IN 47712

INTRODUCTION

The predominant inhibitory function in the brain is a result of the actions of γ-aminobutyric acid (GABA), first described in nerve tissue by Roberts and Frankel (1950). What we now know of the role of GABA in central nervous system activity is reviewed by Krnjevic (this volume). It is believed that GABA is inactivated at the synapse by an active Na^+- and Cl^--dependent transport process responsible for its removal from the synaptic cleft and therefore from the vicinity of adjacent receptors (Kanner, 1978; Pastuszko et al., 1981; Liron et al., 1988; King and Tunnicliff, 1990). After being taken up by presynaptic neurons and glial cells, the intracellular GABA is catabolized by the combined catalytic action of GABA aminotransferase (EC 2.6.1.19) and succinic semialdehyde dehydrogenase (EC 1.2.1.24) which appear to exist as a complex in the matrix of the mitochondria (Hearl and Churchich, 1984). GABA aminotransferase is considered to be the rate-limiting enzyme of the pair (Miller and Pitts, 1967; Buu and Van Gelder, 1974). The reaction catalyzed is the transfer of an amino group from GABA to α-ketoglutarate in the presence of pyridoxal 5'-phosphate as cofactor. The enzyme isolated from pig brain consists of two identical subunits each of approximately 50,000 molecular mass (Churchich and Moses, 1981). Interestingly, Choi and Churchich (1986) have synthesized GABA aminotransferase from polysomal RNA isolated from pig brain and found the relative molecular mass to be 52,000. Despite the obvious importance of GABA aminotransferase, some of the neuronally sequestered GABA might be available for re-release rather than undergoing metabolic degradation (Gram et al., 1988).

Early studies pointed to GABA being involved in certain types of convulsive states in experimental animals. For instance, the inhibition of glutamate decarboxylase by the administration of several drugs led to a reduction in brain GABA concentrations and to the development of generalized seizures (Killam and Bain, 1957). Results from numerous other studies have confirmed that compounds able to reduce brain GABA levels often induce epileptic-like seizures in experimental animals. These observations led to the notion that an elevation of GABA activity in the nervous system might protect against experimentally induced convulsions. There are many reports which detail experiments performed to test this idea and to attempt to develop clinically useful antiepileptic agents. In principle, there are several ways to achieve increased brain GABA function: a) by inhibiting GABA uptake from the synaptic gap (see Chapter 8, this volume), b) by stimulating the $GABA_A$ receptor and thus activating neuronal Cl^- influx (see Chapters 4, 7, and 10, this volume), c) by stimulating GABA biosynthesis and release, and d) by inhibiting the metabolic degradation of GABA. The latter method is the basis of this review.

TYPES OF INHIBITION OF GABA AMINOTRANSFERASE

The four, broad types of inhibitors that will be considered are a) reversible, competitive inhibitors, b) affinity labels, c) mechanism-based inhibitors, and d) residue-specific reagents.

Competitive Inhibitors

Before a compound is administered to an animal in an attempt to increase the concentration of brain GABA it must, of course, first be established that it can inhibit GABA aminotransferase in vitro. The most obvious inhibitors are those that would reversibly compete for binding at the active site—either as structural analogues of substrate (i.e., GABA or α-ketoglutarate) or as structural analogues of pyridoxal 5'-phosphate. These inhibitors (I) form a dead-end complex (EI) with the enzyme (E) which complex can then dissociate to yield the original enzyme and inhibitor:

$$E \; + \; I \; \underset{k_1}{\overset{k_1}{\rightleftharpoons}} \; EI$$

The effectiveness of this type of inhibitor is measured by the inhibition constant (K_i), an equilibrium constant defined as k_{-1}/k_1. The units of K_i are moles/liter; the smaller the K_i value, the more potent the drug. In principle, some drugs could appear to be competitive inhibitors by virtue of their ability to complex with either GABA, α-ketoglutarate or pyridoxal 5'-phosphate. Indeed, many compounds have been identified which react with the cofactor—these are known as carbonyl-trapping agents. Inhibitors that compete with or react with with α-ketoglutarate or pyridoxal 5'-phosphate will obviously not show much selectivity in their action since these two substances are involved in numerous enzymic reactions.

In practice most of the competitive inhibitors of GABA aminotransferase have been found to interfere with either GABA or with the pyridoxal 5'-phosphate. Hydroxylamine was one of the first compounds reported to inhibit the enzyme in vitro. This effect was competitive with respect to GABA with an apparent K_i of about 0.2 mM. When hydroxylamine was administered to mice an increase in brain GABA levels was observed together with a decrease in GABA aminotransferase activity 90 min later (Baxter and Roberts, 1961). Subsequent studies established that hydroxylamine was able to protect against experimentally induced-seizures in animals (Lehmann, 1964; Kohli and Kishor, 1965). Hydroxylamine is able to penetrate the blood-brain barrier and so affect GABA metabolism. Many GABA aminotransferase inhibitors, however, cannot easily enter the central nervous system after parenteral administration and are consequently less valuable in epilepsy research and presumably would be ineffective as therapeutic agents.

One of the most potent inhibitors of GABA aminotransferase is aminooxyacetate (AOAA) which was first demonstrated to inhibit the enzyme from *E. coli* (K_i = 7.5 μM) by competing for GABA binding sites (Wallach, 1961). The rat brain enzyme, however, is about 100 times more sensitive to the effects of the drug (Tunnicliff et al., 1977; Ngo and Tunnicliff, 1978). A contributing factor to AOAA's inhibitory effects appears to be an interference with pyridoxal 5'-phosphate since the inhibitor is a carbonyl-trapping agent and can form an oxime with the cofactor (Roberts and Simonsen, 1963). Many reports attest to the anticonvulsant action of AOAA although it has not been fully established that this effect is a result of its inhibition of GABA aminotransferase (Da Vanzo et al., 1961; Roa et al., 1964; Kuriyama et al., 1966; Meldrum et al., 1970; Osuide, 1972; Wood and Peesker, 1976). It should be pointed out, however, that AOAA can also *cause* seizures under certain conditions and consequently it is obviously of limited use as an anticonvulsant. The mechanism for this convulsant effect is probably related to the drug's ability to also inhibit glutamate decarboxylase under certain conditions (Wood and Peesker, 1973).

Another competitive inhibitor with respect to GABA is 5-ethyl-5-phenylpyrrolidone. This drug has been reported to protect mice against thiosemicarbazide, pentylenetetrazole, and electroshock-induced convulsions (Carvajal et al., 1964; Perez de la Mora and Tapia, 1973). Other structural analogues of GABA include cycloserine which is known to both elevate brain GABA concentrations and to inhibit GABA aminotransferase after intraperitoneal injection (Scotto et al., 1963). Later experiments demonstrated that cycloserine could reduce audiogenic seizures in mice (Dann and Carter, 1964; Lehmann, 1964).

Buu and Van Gelder (1974) showed that both 3-chloroGABA and 3-phenylGABA were weak competitive inhibitors of GABA aminotransferase in vitro. Neither compound, however, was able to increase brain GABA levels after an intraperitoneal injection to mice. Both compounds apparently entered the brain since administration to cats altered EEG activity. 3-PhenylGABA caused the appearance of synchronized, slow-wave activity, whereas 3-chloroGABA had no affect in the normal cat. In epileptic cats, however, 3-chloroGABA greatly reduced or even abolished epileptic EEG discharges.

Hydrazinopropionic acid inhibited mouse brain GABA aminotransferase in a competitive manner with respect to GABA. A K_i value of 2.35×10^{-7}M was reported (Van Gelder, 1968). Subcutaneous injections of the compound (15-30 mg/kg) into mice gave rise to a sedation lasting for several hours. A few animals displayed generalized seizures from which they died. Measurement of GABA levels revealed a gradual increase even up to 16 hours later. GABA aminotransferase activity was completely inhibited. Glutamate decarboxylase was also moderately affected, demonstrating that hydrazinopropionic acid lacks specificity.

For various reasons, none of drugs discussed above has clinical applications. Sodium valproate, on the other hand, is an important clinical agent in the treatment of generalized epileptic seizures (Carraz et al., 1964). The free acid, which is a branched fatty acid and a liquid at room temperature, had previously been shown to protect mice against pentylenetetrazol seizures but the sodium salt proved to be a more convenient form of the drug. There is evidence that valproate owes its anticonvulsant properties to its ability to enhance GABA function in the central nervous system. For instance, administration of the drug to experimental animals leads to an increase in brain GABA concentrations (Godin et al., 1969). Presumably the elevated GABA levels would have to be at nerve endings rather than in glial cells for the effect to have significance. The experiments of Löscher (1981) suggest that this is indeed the case. Following valproate treatment, increased GABA levels were confined to synaptosomal fractions. Moreover, valproate may alter GABA metabolism in humans because long-term drug treatment of epileptic patients and healthy volunteers leads to an increase in GABA levels in the cerebrospinal fluid (Löscher and Schmidt, 1980; Löscher and Siemes, 1984).

How does valproate administration lead to increases in brain GABA levels? The drug has been shown to weakly inhibit GABA aminotransferase isolated from brain by acting as a competitive inhibitor with respect to GABA (Godin et al., 1969; Fowler et al., 1975). Inhibitor constants of between 10 mM and 100 mM have been reported (Chapman et al., 1982). Valproate, on the other hand, also inhibits other enzymes involved in GABA metabolism, i.e., succinic semialdehyde dehydrogenase and aldehyde reductase (Harvey et al., 1975; Van der Laan, 1979). Interestingly, several studies have failed to demonstrate that valproate administration leads to inhibition of GABA aminotransferase in vivo (Löscher and Nau, 1982; Nau and Löscher, 1982). Another mechanism contributing to an elevation of brain GABA could be an increase in glutamate decarboxylase activity induced by valproate (Chapman et al., 1982; Phillips and Fowler, 1982; Löscher, 1989).

In addition to its apparent effects on GABA degradation, electrophysiological studies have shown that valproate also augments the inhibitory effects of GABA on neuronal activity (Schmultz et al., 1979; Macdonald and Bergey, 1979; Baldino and Geller, 1981; Preisendorfer et al., 1987). Since these studies required comparatively high concentrations of valproate, the potentiation of GABA's action may not have significance in explaining the antiepileptic properties of valproate. Another interesting observation is that valproate inhibits the binding of [^{3}H] dihydropicrotoxinin to the GABA$_A$ receptor complex (Ticku and Davis, 1981). Again, however, comparatively high concentrations of drug are required for this effect. In addition, Koe (1983) noted that valproate enhanced [^{3}H] flunitrazepam binding to mouse brain particles. A further effect that valproate has on GABA neurotransmission that is not directly related to GABA aminotransferase inhibition, is an enhanced GABA turnover in the substantia nigra (Löscher, 1989).

Before the idea of a GABA involvement in the anticonvulsant mechanism of valproate action can be accepted, several key questions must be answered satisfactorily: 1) Do clinically efficacious doses of valproate elevate GABA concentrations in the human brain? 2) Are the sites of such increases neuronal? 3) If valproate potentiates neuronal inhibition by GABA, are the concentrations of the drug in the brain high enough for this to be a contributing mechanism in the antiepileptic mechanism of valproate? Unfortunately, no direct evidence exists that GABA is increased in the brain of epileptics treated with valproate. Furthermore, in some studies no increase in CSF GABA has been demonstrated. Other mechanisms related to the anticonvulsant action of valproate have been explored. For example, a correlation was reported between the ability of a series of valproate analogues to protect against seizures and their ability to decrease brain aspartate concentrations in rodents (Schechter et al., 1978). In addition, valproate can inhibit voltage-dependent Na$^+$ channels in cultured neurons (McLean and Macdonald, 1986). Obviously it is possible that a combination of mechanisms are at the root of valproate's clinical effects.

The amino acid gabapentin [1-(aminomethyl)cyclohexane acetic acid] was designed to resemble the steric confirmation of GABA and to penetrate the blood-brain barrier. It does indeed penetrate the brain where it exhibits a broad spectrum of anticonvulsant activity (Ojemann et al., 1988). Perhaps surprisingly, gabapentin has not been shown to have an affinity for the GABA$_A$ receptor (Schmidt, 1989). The drug, however, does have a weak inhibitory effect on GABA aminotransferase similar to valproate. Whether or not this inhibitory action accounts for its anticonvulsant activity is unknown.

Another antiepileptic drug currently undergoing clinical trials is stiripentol (4,4-dimethyl-1-[(3,4 methylenedioxy)phenyl]-1-penten-3-ol). This agent has a broad spectrum of anticonvulsant activity. For instance, it is effective against both electrically- and chemically-induced seizures (Poisson et al., 1984). In mice stiripentol increases whole-brain GABA levels, probably as a result of GABA aminotransferase inhibition (Wegmann et al., 1978).

Clifford et al. (1973) measured the effects of imidazoleacetic acid on mouse brain GABA aminotransferase activity in vitro. Although this substance bears a close structural resemblance to GABA, kinetic analysis demonstrated that it inhibited GABA aminotransferase in a noncompetitive manner. This inhibitor, known to be a potent GABA$_A$ receptor agonist, has recently been detected in low concentrations in rat brain and human cerebrospinal fluid (Khandelwal et al., 1989). It has not been established that imidazoleacetic acid can inhibit GABA aminotransferase in vivo even though it can pass through the blood-brain barrier at relatively high doses. In fact, following intraperitoneal administration (3 mmole/kg) of IMA, no alterations in brain GABA levels were seen even up to 2 hours later (Tunnicliff et al., 1972). Treatment of mice with IMA reportedly causes a hypnotic state together with analgesia (Roberts and Simonsen, 1966). Despite the induction of "sleep," IMA produces a strong CNS excitatory activity (Marcus et al., 1971) and can even induce tonic-clonic seizures in mice (Tunnicliff and Lucas, unpublished observations).

Finally, compounds could, in principle, possess anticonvulsant properties by inhibiting GABA aminotransferase as a consequence of interfering with pyridoxal 5'-phosphate (see review by Tunnicliff, 1989). However, since pyridoxal 5'-phosphate is required as a cofactor in numerous biochemical reactions, it would be difficult to design a specific inhibitor of GABA aminotransferase by this approach.

Inactivators

Affinity labels. An inactivation is the irreversible inhibition of an enzyme. As with reversible inhibition, inactivation can also be of a competitive nature. This is because often an irreversible inhibitor will initially compete with substrate for binding at the active site. However, once at the active site the inactivator forms covalent bonds with neighboring amino acid residues. These inhibitors are usually called affinity labels. The overall reaction can be described thus:

$$E + I \underset{k_{-1}}{\overset{k_1}{\rightleftharpoons}} EI \xrightarrow{k_2} E - I$$

The effectiveness of affinity labels is measured by the inactivation constant (K_{inact}), the concentration of drug that produces a 50% maximum rate of inactivation. This constant defined as $k_{-1} + k_2/k_1$. Because the irreversible covalent bond formation step is usually rate-limiting, it can be seen that the K_{inact} is very similar to the dissociation constant for the reversible inhibitor-enzyme interaction. Thus the K_{inact} and the K_i for a reversible competitive inhibitor are comparable terms.

Several fluorinated derivatives of GABA have been shown to inactivate pig brain GABA aminotransferase in a time-dependent manner (Bey et al., 1981; Schirlin et al., 1987). The mechanism proposed is an enzyme catalyzed removal of a proton alpha to the imine formed from pyridoxal 5'-phosphate and the inhibitor, followed by elimination of a fluoride ion. The most effective inactivators in vitro were 3-amino-4-fluorobutanoic acid, 3-amino-4, 4-difluorobutanoic acid, 4-amino-5-fluoropentanoic acid and 3-amino-2,

4-difluorobutanoic acid. These all increased brain GABA levels and inhibited GABA aminotransferase after intraperitoneal administration to mice. The most potent of these inhibitors was 3-amino-4, 4-difluorobutanoic acid, which exhibited an IC_{50} of about 2mg/kg.

Schirlin et al. (1987) synthesized three halogen derivatives of GABA and reported that each displayed irreversible inhibition of GABA aminotransferase purified from pig brain. 3-Amino-4, 4-difluorobutanoic acid, 3-amino-4-chloro-4-fluorobutanoic acid and 3-amino-2, 4-difluorobutanaoic acid each inactivated the enzyme with K_{inact} values in the low millimolar range.

Mechanism-based inhibitors. Some inactivators are not reactive enough to form covalent bonds at the active site of an enzyme but can be converted by the enzyme to a reactive species. These compounds act as substrates and consequently often have a high affinity for the active site. An inactivation constant (K_I) can be calculated which is a measure of the potency of these inhibitors. This constant is different from the K_{inact} for the affinity labels:

$$E + I \underset{k_{-1}}{\overset{k_1}{\rightleftharpoons}} EI \xrightarrow{k_2} E - I$$

$$EI \xrightarrow{k_3} E + I$$

Thus the $K_I = K_{inact}[(k_3 + k_4)/k_2 + k_3 + k_4)]$

The first reported mechanism-based inhibitor was ethanolamine-*O*-sulphate which Fowler and John (1972) found to be an irreversible competitive inhibitor of GABA aminotransferase purified from rabbit brain, GABA being able to protect against the initial binding. The K_I was reported as 0.44 µM. This drug is not able to penetrate the blood-brain barrier, but when administered intracerebrally to mice it exerted an anticonvulsant effect against electroconvulsive shock and sound-induced seizures (Baxter et al., 1973; Anlezark et al., 1976). Recently Nanavati and Silverman (1989) proposed that mechanism-based inhibitors of GABA aminotransferase could be divided into four classes depending upon their mechanism of action (see Table I). Ethanolamine-*O*-sulfate was put in Class II because its inactivation of the enzyme involves a β-elimination step, the formation of an enamine intermediate and the subsequent formation of a ternary adduct of an active site lysine residue, pyridoxal 5′-phosphate, and the inactivator. Much of the elucidation of the mechanism of ethanolamine-*O*-sulfate derives from the experiments of Likos et al. (1982) and Ueno et al. (1982).

Class I inactivators are those producing an electrophile that by conjugate addition attaches to an essential amino acid residue. The best known example of this type of inhibitor is γ-vinylGABA (vigabatrin or 4-amino-5-hexenoic acid). This drug is a very effective anticonvulsant in several models of epilepsy. Examples include photogenic convulsions in baboons (Meldrum and Horton, 1978), audiogenic seizures in mice (Schechter et al., 1977), and amygdala-kindled seizures (Myslobodsky et al., 1979). Clinical trials have established that vigabatrin has a wide spectrum of activity including control of complex partial seizures (Gram et al., 1983; Remy et al., 1986; Rimmer and Richens, 1984; Tartara et al., 1986; Tassinari et al., 1987; Dam, 1989). Vigabatrin readily crosses the blood-brain barrier and raises brain GABA levels by vitue of GABA aminotransferase inhibition (Jung et al., 1977). The decrease in GABA aminotransferase activity and the increase in brain GABA concentration lasts for several days after a single dose of the drug (Lippert et al., 1977; Larsson et al., 1986). Moreover, evidence indicates that it is the elevation of GABA in the substantia nigra that is important in its mechanisms of action (Gale and Iadarola, 1980; Iadarola and Gale, 1982; Gale, 1989). Several studies have shown that administration of vigabatrin to patients leads to an increase in GABA levels in the CSF (e.g., Riekkinen et al., 1989).

Another mechanism-based inhibitor is gabaculline, first isolated from *Streptomyces toyocaensis*. This substance was reported to inactivate mouse brain GABA aminotransferase (K_I = 0.6 μM) (Rando and Bangerter, 1976). The closely related compound isogabaculline has a similar effect on the enzyme (Rando, 1977). Both these inactivators belong in Class III. These are compounds which inactivate GABA aminotransferase by forming stable

TABLE I. Criteria for Classification of Mechanism-Based Inhibitors of GABA Aminotransferase and Examples for Each Class

Class	Drug	Reference Criteria
I	Vigabatrin	1. Covalently binds with active-site residue
II	Ethanolamine-O-sulfate	2. Enamine formation leads to a ternary adduct with cofactor and lysine residue
III	Gabaculline	3. Forms adduct with cofactor not attached to lysine residue
IV	5-Fluoro-4-oxopentanoic acid	4. Reacts with the pyridoxamine phosphate form of the enzyme

Adapted from Nanavati and Silverman (1989).
1. Lippert et al. (1977).
2. Fowler and John (1972).
3. Rando and Bangerter (1976).
4. Lippert et al. (1982).

adducts with the cofactor pyridoxal 5′-phosphate, rather than covalently binding to active-site residues.

Although 3-nitro-1-propanamine was unable to inhibit GABA aminotransferase purified from pig brain, it did inactivate the enzyme from *Pseudomonas fluorescens*. The homologue 4-nitro-1-butanamine, on the other hand, has been reported to act as an irreversible inhibitor of the brain enzyme, even though no details were given (Alston et al., 1987). According to Nanavati and Silverman (1989) both these compounds belong to Class III, although there is some doubt that the normal enzymic catalytic mechanism is involved.

Fluoro-4-ketopentanoic acid was reported to act as a mechanism-based inhibitor (Lippert et al., 1982) and appeared to fall into a Class I mechanism. But in light of subsequent work by Likos et al. (1982) and Ueno et al. (1982), it is now thought this agent reacts with the pyridoxamine 5′-phosphate at the active site, with the eventual formation of an enamine. Consequently this inhibitor has been classified as belonging to Class IV, i.e., any inactivator which requires the pyridoxamine 5′-phosphate form of an enzyme (Nanavati and Silverman, 1989). The only other Class IV inactivator so far described is 3,5-dioxocyclohexane-carboxylic acid, a structural analogue of succinic semialdehyde, which has been shown to inactivate GABA aminotransferase from both *Pseudomonas fluorescens* and pig brain (Alston et al., 1982). No inactivation rates were calculated nor were K_I values given. Table I summarizes the criteria necessary for an inhibitor to belong any one of the four classes of GABA aminotransferase inactivators.

Residue-specific inhibitors.   Compounds which react fairly specifically with essential amino acids of a protein but which do not have a particular affinity for the active site have proven to be extremely valuable in elucidating residues that play a role in the catalytic mechanism (Means and Feeney, 1971). Such chemicals produce their irreversible inhibition by reacting with the enzyme by a nonspecific bimolecular reaction displaying second order kinetics. Accordingly, the existence of an EI complex is not measurable and a K_{inact} cannot be calculated:

$$E \;+\; I \quad\xrightarrow{\;\;k_2\;\;}\quad E \;-\; I$$

In the case of GABA aminotransferase, the sulfhydryl group reagents, 5,5′-dithiobis-2-nitrobenzoate and N-iodoacetylaminoethyl-5-naphthylamine sulfonate have both been shown to inactivate the enzyme (Moses and Churchich, 1980), suggesting that cysteine residues are necessary for enzyme activity.

Phenylglyoxal has been employed extensively to investigate the role of arginine residues in the catalytic mechanism of many enzymes (Riordan,

1979). This chemical inactivates mouse brain GABA aminotransferase, an effect which is reduced in the presence of pyridoxal 5′-phosphate, indicating the involvement of essential arginine residues at the cofactor binding site (Tunnicliff, 1980).

Inactivators which form covalent bonds with essential amino acids via a bimolectular reaction probably have no direct pharmacological application, mainly because their lack of potency and specificity. However, from studies employing other residue-specific inhibitors it is evident that lysine residues are located at or near the α-ketoglutarate binding site on pig brain GABA aminotransferase (Kim and Churchich, 1981; Kim and Churchich, 1987). Such knowledge of the active site of GABA aminotransferase could well assist in the development of compounds able to inhibit the enzyme and possibly act as anticonvulsants.

CONCLUSIONS

This chapter has attempted to explore the different types of mechanisms that can account for GABA aminotransferase inhibition. Evidence is available that only two of these mechanisms are involved in the anticonvulsant action of clinically relevant drugs. That is, a competitive, reversible inhibition which could be, at least in part, the basis of the action of sodium valproate; and the mechanism-based inactivation of GABA aminotransferase by vigabatrin.

For some reason, anticonvulsants like vigabatrin that are potent inhibitors of GABA aminotransferase have fewer side effects (sedation, proconvulsant effects and ataxia, for example) than drugs that enhance GABA function by other mechanisms (GABA uptake inhibitors, for instance). Meldrum (1989) has suggested that perhaps a desensitization to GABA may occur with GABA agonists, an effect not seen with GABA aminotransferase inhibitors. With the latter, an enhanced release of transmitter probably gives a more effective suppression of seizures rather than a flooding of the synapic cleft with inhibitory agent. Although some of the most effective antiepileptic agents are those that enhance GABA neurotransmission by acting at the GABA/benzodiazepine receptor complex, drugs like vigabatrin have demonstrated that the inhibition of GABA aminotransferase is a viable alternative mechanism for useful clinical anticonvulsants. There is no reason to expect that other medications effective in treating epilepsy by virtue of brain GABA aminotransferase inhibition will not be developed.

REFERENCES

Alston TA, Porter DJT, Bright HJ: Inactivation of GABA aminotransferase by 3-nitro-1-propanamine. J Enzym Inhib 1: 215–222, 1987.

Alston TA, Porter DJT, Wheeler MS, Bright HJ: Mechanism-based inactivation of GABA aminotransferase by 3,5-dioxocyclohexanecarboxylic acid. Biochem Pharmacol 31: 4081–4084, 1982.

Anlezark G, Horton RW, Meldrum BS, Sawaya MC: Anticonvulsant action of ethanolamine-*O*-sulphate and di-*n*-propylacetate and the metabolism of γ-aminobutyric acid (GABA) in mice with audiogenic seizures. Biochem Pharmacol 25: 413–417, 1976.

Baldino F, Geller HM: Sodium valproate enhancement of GABA inhibition: Electrophysiological evidence for anticonvulsant activity. J Pharmacol Exp Ther 217: 445–450, 1981.

Baxter MG, Fowler LJ, Miller AA, Walker JMG: Some behavioural and anticonvulsant actions in mice of ethanolamine O-sulphate, an inhibitor of 4-aminobutyrate aminotransferase. Br J Pharmacol 47: 681P, 1973.

Baxter CF, Roberts E: Elevation of γ-aminobutyric acid in brain: selective inhibition of γ-aminobutyric-α-ketoglutaric acid transaminase. J Biol Chem 236, 3287–3294, 1961.

Bey P, Jung MJ, Gerhart F, Schirlin D, Van Dorsselaer V, Casara P: ω-Fluoromethyl analogues of ω-amino acids as irreversible inhibitors of 4-aminobutyrate: 2-oxoglutarate aminotransferase. J Neurochem 37: 1341–1344, 1981.

Burkhart JP, Holbert GW, Metcalf BW: Enantiospecific synthesis of (S)-4-amino-4, 5-dihydro-2-furancarboxylic acid, a new suicide inhibitor of GABA-transaminase. Tetrah Letts 25: 5267–5270, 1984.

Buu NT, Van Gelder NM: Biological actions in vivo and in vitro of two γ-aminobutyric acid (GABA) analogues: γ-ChloroGABA and γ-phenylGABA. Br J Pharmacol 52: 401–406, 1974.

Carraz G, Farr R, Chateau R, Bonnin J: First clinical trials of the antepileptic activity of n-dipropylacetic acid. Ann Med Psychol (Paris) 122: 577–584, 1964.

Carvajal G, Russek M, Tapia R, Massieu G: Anticonvulsive action of substances designed as inhibitors of γ-aminobutyric acid-α-ketoglutaric acid transaminase. Biochem Pharmacol 13: 1059–1069, 1964.

Chapman A, Keane PE, Meldrum BS, Simiand J, Verniers JC: Mechanism of anticonvulsant action of valproate. Prog Neurobiol 19: 315–359, 1982.

Choi S-Y, Churchich JE: Biosynthesis of 4-aminobutyrate aminotransferase. Eur J Biochem 161: 289–294, 1986.

Churchich JE, Moses U: 4-Aminobutyrate aminotransferase: The presence of nonequivalent binding sites. J Biol Chem 256, 1101–1104, 1981.

Clifford JM, Tabener PV, Tunnicliff G, Rick JT, Kerkut GA: Biochemical and pharmacological actions of imidazoleacetic acid. Biochem Pharmacol 22: 535–542, 1973.

Da Vanzo JP, Greig ME, Cronin MA: Anticonvulsant properties of amino-oxyacetic acid. Am J Physiol 201, 833–837, 1961.

Dam M: Long-term evaluation of vigabatrin (gamma-vinylGABA) in epilepsy. Epilepsia 30 (suppl. 30), S26–S30, 1989.

Dann OT, Carter CE: Cycloserine inhibition of gamma-aminobutyric-alpha-ketoglutaric transaminase. Biochem Pharmacol 13: 677–684, 1964.

Fowler LF, John RA: Active-site-directed irreversible inhibition of rat brain 4-aminobutyrate aminotransferase by ethanolamine O-sulphate in vitro and in vivo. Biochem J 130: 569–573, 1972.

Fowler LJ, Beckford J, John RA: An analysis of the kinetics of the inhibition of rabbit brain aminobutyrate aminotransferase by sodium n-dipropylacetate and some other simple carboxylic acids. Biochem Pharmacol 24: 1267–1270, 1975.

Gale K: Mechanisms of seizure control mediated by γ-aminobutyric acid: role of the substantia nigra. Fed Proc 44: 2414–2424, 1985.

Gale K: GABA in epilepsy: The pharmacological basis. Epilepsia 30 (suppl. 30), S1–S11, 1989.

Gale K, Iadarola MJ: Seizure protection and increased nerve-terminal GABA: Delayed effects of GABA transaminase inhibition. Science 208: 288–291, 1980.

Godin Y, Heiner L, Mark J, Mandel P: Effects of di-*n*-propylacetate, an anticonvulsant compound, on GABA metabolism. J Neurochem 16, 869–873, 1969.

Gram L: Experimental studies and controlled clinical testing of valproate and vigabatrin. Acta Neurol Scand 78, 241–270, 1988.

Gram L, Lyon BB, Dam M: Gamma-vinyl-GABA: A single blind trial in patients with epilepsy. Acta Neurol Scand 68, 34–39, 1983.

Gram L, Larsson OM, Johnsen AH, Schousboe A: Effects of valproate, vigabatrin and aminooxyacetic acid on release of endogenous and exogenous GABA from cultured neurons. Epilepsy Res 2: 87–95, 1988.

Harvey PKP, Bradford HF, Davison AN: The inhibitory effect of sodium n-dipropylacetate on the degradative enzymes of the GABA shunt. FEBS Lett 52, 251–254, 1975.

Hearl WG, Churchich JE: Interactions between 4-aminobutyrate aminotransferase and succinic semialdehyde dehydrogenase, two mitochondrial enzymes. J Biol Chem 259: 11459–11463, 1984.

Iadarola MJ, Gale K: Substantia nigra: site of anticonvulsant activity mediated by gamma-aminobutyric acid. Science 218: 1237–1240, 1982.

Jung MJ, Lippert B, Metcalf BW, Bohlen P, Schechter PJ: γ-Vinyl GABA (4-amino-hex-5-enoic acid), a new selective irreversible inhibitor of GABA-T: Effects on brain GABA metabolism in mice. J Neurochem 29, 797–802, 1977.

Kanner BI: Active transport of γ-aminobutyric acid by membrane vesicles isolated from rat brain. Biochemistry 17: 1207–1211, 1978.

Khandelwal JK, Prell GD, Morrishow AM, Green JP: Presence and measurement of imidazoleacetic acid, a γ-aminobutyric acid agonist, in rat brain. J Neurochem 52: 1107–1113, 1989.

Killam KF, Bain JA: Convulsant hydrazides I. In vitro and in vivo inhibition of vitamin B_6 enzymes by convulsant hydrazides. J Pharmacol Exp Ther 119: 255–262, 1957.

Kim DS, Churchich JE: 4-Aminobutyrate aminotransferase, the reaction of lysine residues connected with enzymatic activity. Biochem Biophys Res Commun 99: 1333–1340, 1981.

Kim DS, Churchich JE: Active site modification of 4-aminobutyrate aminotransferase with ATP analogs. Biochim Biophys Acta 916: 265–270, 1987.

King SM, Tunnicliff G: Na^+- and Cl^--dependent [^{3}H]GABA binding to catfish brain particles. Biochem Int 20: 821–831, 1990.

Koe BK: Enhancement of benzodiazepine binding by progabide (SL 76002) and SL 75102. Drug Dev Res 3: 421–432, 1983.

Kohli RP, Kishor K: The anticonvulsant activity of hydroxylamine. Arch Int Pharmacodyn Ther 154: 89–93, 1965.

Kuriyama K, Roberts E, Rubinstein MK: Elevation of γ-aminobutyric acid in brain with aminooxyacetic acid and susceptibility to convulsive seizures in mice: A quantitative re-evaluation. Biochem Pharmacol 15: 221–236, 1966.

Larsson OM, Gram L, Schousboe I, Schousboe A: Differential effects of gamma-vinylGABA and valproate on GABA-transaminase from cultured neurons and astrocytes. Neuropharmacology 25: 617–625, 1986.

Lehmann A: Contribution a l'etude psychophysiologique et neuropharmacologique de l'epilepsie acoustique de la souris et du rat. II. Etude experimental. Aggressologie 5: 311–351, 1964.

Likos JJ, Ueno H, Feldhaus RW, Meltzer DE: A novel reaction of the coenzyme of glutamate decarboxylase with L-serine *O*-sulfate. Biochemistry 21: 4377–4386, 1982.

Lippert B, Metcalf BW, Jung MJ, Casara P: 4-Amino-hex-5-enoic acid, a selective catalytic inhibitor of 4-aminobutyric acid aminotransferase in mammalian brain. Eur J. Biochem 74: 441–445, 1977.

Lippert B, Metcalf BW, Resvick RJ: Enzyme-activated irreversible inhibition of rat and mouse brain 4-aminobutyric acid-α-ketoglutarate transaminase by 5-fluoro-4-oxopentanoic acid. Biochem Biophys Res Commun 108, 146–152, 1982.

Liron Z, Wong E, Roberts E: Studies on uptake of γ-aminobutyric acid by mouse brain particles; toward the development of a model. Brain Res 444, 119–132, 1988.

Löscher W: Valproate-induced changes in GABA metabolism at the subcellular level. Biochem Pharmacol 30: 1364–1366, 1981.

Löscher W: Valproate enhances GABA turnover in the substantia nigra. Brain Res 501: 198–203, 1989.

Löscher W, Nau H: Valproic acid: metabolite concentrations in plasma and brain, anticonvulsant activity, and effects on GABA metabolism during subacute treatment in mice. Arch Int Pharmacodyn Ther 257: 20–31, 1982.

Löscher W, Siemes H: Valproic acid increases GABA in CSF of epileptic children. Lancet II, 225, 1984.

Löscher W, Schmidt D: Increase of human plasma GABA by sodium valproate. Epilepsia 21, 611–615, 1980.

Macdonald RL, Bergey GK: Valproic acid augments GABA-mediated postsynaptic inhibition in cultured mammalian neurons. Brain Res 170: 558–562, 1979.

Marcus RJ, Winters WD, Roberts E, Simonsen DG: Neuropharmacological studies of imidazole-4-acetic acid actions in the mouse and rat. Neuropharmacology 10: 203–215, 1971.

McLean MJ, Macdonald RL: Sodium valproate, but not ethosuximide, produces use- and voltage dependent limitation of high frequency repetitive firing of action potentials of mouse central neurons in cell culture. J Pharmacol Exp Ther 237: 1001–1011, 1986.

Means GE, Feeney RE: Chemical Modification of Proteins. San Francisco: Holden-Day, Inc., 1971.

Meldrum BS: GABAergic mechanisms in the pathogenesis and treatment of epilepsy. Br J Clin Pharmacol 27: 3S–11S, 1989.

Meldrum BS, Balzano E, Gadea M, Naquet R: Photic and drug-induced epilepsy in the baboon (Papio papio): the effects of isoniazid, thiosemcarbazide, pyridoxine and amino-oxyacetic acid. Electroencephalogr Clin Neurophysiol 29: 333–347, 1970.

Meldrum BS, Horton R: Blockade of epileptic responses in the photosensitive baboon, Papio papio, by two irreversible inhibitors of GABA transaminase, γ-acetylenic GABA (4-amino-hex-5-ynoic acid) and γ-vinyl GABA (4-amino-hex-5-enoic acid). Psychopharmacology (Berl) 59: 47–50, 1978.

Miller AL, Pitts FN: Brain succinic semialdehyde dehydrogenase - II. Activities in twenty-four regions of human brain. J Neurochem 14: 579–584, 1967.

Moses U, Churchich JE: Reactivity of one thiol group in the dimeric protein, 4-aminobutyrate aminotransferase. Biochim Biophys Acta 613: 392–400, 1980.

Myslobodsky MS, Ackermann RF, Engel J: Effects of γ-acetylenic GABA and γ-vinyl GABA on metrozol-activated and kindled seizures. Pharmacol Biochem Behav 11: 265–271, 1979.

Nanavati SM and Silverman RB: Design of potential anticonvulsant agents: Mechanistic classification of GABA aminotransferase inactivators. J Med Chem 32: 2413–2421, 1989.

Nau H, Löscher W Valproic acid: Brain and plasma levels of the drug and its metabolites, anticonvulsant effects and γ-aminobutyric acid (GABA) metabolism in the mouse. J Pharmacol Exp Ther 220: 654–659, 1982.

Ngo TT, Tunnicliff G: Further evidence for the existence of isozymes of brain γ-aminobutyrate aminotransferase. Comp Biochem Physiol 59C: 101–104, 1978.

Ojemann LM, Friel PN, Ojemann GA: Gabapentin concentration in human brain. Epilepsia 29: 694, 1988.

Osuide G: Pharmacological properties of aminooxyacetic acid in the chicken. Br J Pharmacol 44: 31–44, 1972.

Pastuszko A, Wilson DF, Erecinska M: Net uptake of γ-aminobutyric acid by a high affinity system of rat brain synaptosomes. Proc Nat Acad Sci USA 78: 1242–1244, 1981.

Perez de la Mora M, Tapia R: Anticonvulsant effect of 5-ethyl, 5-phenyl, 2-pyrrolidone and its possible relationship to gamma-aminobutyric acid-dependent inhibitory mechanisms. Biochem Pharmacol 22: 2635–2639, 1973.

Phillips NI, Fowler LJ: The effects of sodium valproate on gamma-aminobutyric acid metabolism and behaviour in naive and ethanolamine-O-sulphate pretreated rats and mice. Biochem Pharmacol 31: 2257–2261, 1982.

Poisson M, Guguet F, Savattier A, Bakri-Logeais F, Narcisse G: A new type of anticonvulsant, stiripentol: pharmacological profile and neurochemical study. Arzneim Forsch Drug Res 34: 199–204, 1984.

Preisendorfer U, Zeise ML, Klee MR: Valproate enhances inhibitory postsynaptic potentials in hippocampal neuron in vitro. Brain Res 435: 213–219, 1987.

Rando RR: Mechanism of the irreversible inhibition of γ-aminobutyric acid α-ketoglutaric acid transaminase by the neurotoxin gabaculline. Biochemistry 16: 4604–4610, 1977.

Rando RR, Bangerter FW: The irreversible inhibition of mouse brain GABA transaminase by gabaculline. J Am Chem Soc 98: 6762–6764, 1976.

Remy C, Favei P, Tell G: Harenberg J, Schechter PJ: Double-blind, placebo-controlled, cross-over study of GVG in drug resistant epilepsy of the adult. Boll Lega It Epil 54/55: 241–243, 1986.

Riekkinen PJ, Pitkanen A, Ylinen A, Sivenius J, Halonen T: Specificity of vigabatrin for the GABAergic system in human epilepsy. Epilepsia 30 (suppl. 3): S18–S22, 1989.

Rimmer EM; Richens A: Double-blind study of gamma-vinylGABA in patients with refractory epilepsy. Lancet 1, 189–190, 1984.

Riordan JF: Arginyl residues and anion binding sites in proteins. Mol Cell Biochem 26: 71–91, 1979.

Roa PD, Tews JR, Stone WE: A neurochemical study of thiosemicarbazide seizures and their inhibition by aminooxyacetic acid. Biochem Pharmacol 13: 477–487, 1964.

Roberts E, Frankel S: -Aminobutyric acid in brain: Its formation from glutamic acid. J Biol Chem 187: 55–63, 1950.

Roberts E, Simonsen DG: Some properties of L-glutamic decarboxylase in mouse brain. Biochem Pharmacol 12: 113–134, 1963.

Roberts E, Simonsen DG: A hypnotic and possible analgesic effect of imidazoleacetic acid in mice. Biochem Pharmacol 15: 1875–1877, 1966.

Schechter PJ, Tranier Y, Jung MJ, Bohlen P: Audiogenic seizure protection by elevated brain GABA concentrations in mice: effects of γ-acetylenic GABA and γ-vinyl GABA, two irreversible GABA-T inhibitors. Eur J Pharmacol 45: 319–328, 1977.

Schechter PJ, Tranier Y, Grove J: Effects of n-dipropylacetate on amino acid concentration in mouse brain: correlation with anticonvulsant activity. J Neurochem 31: 1325–1327, 1978.

Schirlin D, Baltzer S, Heydt JG, Jung MJ: Irreversible inhibition of GABA-T by halogenated analogues of β-alanine. J Enzym Inhib 1: 243–258, 1987.

Schmidt B: Potential antiepileptic drugs: Gabapentin. In Levy RH, Dreifuss FE, Mattson RH, Meldrum BS, Penry JK (eds.): "Antiepileptic Drugs, 3rd Edition." New York: Raven Press, 1989, pp 925–935.

Schmultz M, Olpe HR, Koella WP: Central actions of valproate sodium. J Pharm Pharmacol 31: 413–414, 1979.

Scotto P, Monaco P, Scardi V, Bonavita V: Neurochemical studies with L-cycloserine, a central depressant agent. J Neurochem 10: 831–839, 1963.

Silverman RB, George C: Inactivation of γ-aminobutyric acid aminotransferase by (Z)-4-amino-2-fluorobut-2-enoic acid. Biochemistry 27: 3285, 1988.

Silvermann RB, Levy MA: Irreversible inactivation of pig brain γ-aminobutyric acid-α-ketoglutarate transaminase by 4-amino-5-halopentanoic acids. Biochem Biophys Res Commun 95: 250–255, 1980.

Tartara A, Manni R, Galimberti CA, Hardenberg J, Orwin J, Perruca E: Vigabatrin in the treatment of epilepsy: a double-blind, placebo-controlled study. Epilepsia 27: 717–723, 1986.

Tassinari CA, Michelucci R, Ambrosetto G, Salvi F: Double-blind study of vigabatrin in the treatment of drug-resistant epilepsy. Arch Neurol 44: 907–910, 1987.

Ticku MK, Davis WC: Effect of valproic acid on ^{3}H-diazepam and ^{3}H-dihydropicrotoxinin binding sites at the benzodizepine-GABA receptor ionophore complex. Brain Res 223: 218–222, 1981.

Tunnicliff G: Essential arginine residues at the pyridoxal phosphate binding site of brain γ-aminobutyrate aminotransferase. Biochem Biophys Res Commun 97: 160–165, 1980.

Tunnicliff G: Inhibitors of brain GABA aminotransferase. Comp Biochem Physiol 93A: 247–254, 1989.

Tunnicliff G, Ngo TT, Rojo-Ortega JM, Barbeau A: The inhibition by substrate analogues of γ-aminobutyrate aminotransferase from mitochondria of different subcellular fractions of rat brain Biochem Cell Biol. 55: 479–484, 1977.

Tunnicliff G, Wein J, Roberts E: Effects of imidazoleacetic acid on brain amino acids and body temperature in mice. J Neurochem 19: 2017–2023, 1972.

Ueno H, Likos JJ, Meltzer DE: Chemistry of the inactivation of cytosolic aspartate aminotransferase by serine O-sulfate. Biochemistry 21: 4387–4393, 1982.

Van der Laan JW, De Boer T, Bruinvels J: Di-n-propylacetate and GABA degradation. Preferential inhibition of succinic semialdehyde dehydrogenase and indirect inhibition of GABA transaminase. J Neurochem 32: 1769–1780, 1979.

Van Gelder NM: Hydrazinopropionic acid: A new inhibitor of γ-aminobutyrate transaminase and glutamate decarboxylase. J Neurochem 15: 747–757, 1968.

Wallach DP: Studies on the GABA pathway-I. The inhibition of γ-aminobutyric acid-α-ketoglutaric acid transaminase in vitro and in vivo by U-7524 (amino-oxy-acetic acid). Biochem Pharmacol 5: 323–331, 1961.

Wegmann R, Ilies A, Aurousseau M: Pharmaco-cellular enzymology of the mode of action of stiripentol during the cardiazolic epilepsy. III. The metabolism of lipids, proteins, nucleoproteins and proteoglycans. Cell Mol Biol 23: 455–480, 1978.

Wood JD, Peesker SJ: The role of GABA metabolism in the convulsant and anticonvulsant actions of amino-oxyacetic acid. J Neurochem 20: 379–387, 1973.

Wood JD, Peesker SJ: A dual mechanism for the anticonvulsant action of aminooxyacetic acid. Can J Physiol Pharmacol 54: 534–540, 1976.

10
The GABA/Benzodiazepine Receptor and Future Prospects for GABA-Mimetics in Epilepsy

BRIAN S. MELDRUM

Institute of Psychiatry, Department of Neurology, London SE5 8AF, England

INTRODUCTION

The key basic observations relating to the anticonvulsant properties of GABA agonists, GABA-uptake inhibitors, and GABA-transaminase inhibitors were all established in animal models before 1980. The last 10 years have provided some refinement of detail and have allowed a preliminary evaluation of the clinical relevance of the animal data. The only area in which there have been significant advances likely to alter the direction of future research, is in understanding the molecular biology of the GABA/benzodiazepine (BZ) receptor. This new understanding will have a major impact on research on GABA mechanisms in epilepsy during the next decade, and provides totally novel tools for the discovery of anticonvulsant drugs. The key details in the molecular biology of the GABA/BZ receptor and their implications for future research will be summarised first. Subsequently the prospects for novel agents in the categories of GABA agonist, GABA uptake inhibitor and GABA-transaminase inhibitor will be briefly reviewed.

GABA$_A$/BENZODIAZEPINE RECEPTOR

The anticonvulsant action of 1,4- and 1,5-benzodiazepines depends primarily on enhancement of the inhibitory effect of GABA at receptors linked to GABA receptors that are sensitive to inhibition by bicuculline. The effect of benzodiazepines is to increase the frequency of channel openings induced by a given concentration of GABA. There is a strong correlation between binding affinity in a series of benzodiazepine compounds (as judged by displacement of flunitrazepam from brain membrane preparations) and potency against threshold (clonic) pentylenetetrazol seizures in mice. The same mechanism is probably responsible for the antiepileptic action of benzodiazepines given chronically to treat *absence* attacks or myoclonic seizures in man. A different mechanism probably involving enhanced inactivation of the Na$^+$ channel may be responsible for the suppression of *Status epilepticus* by high doses of diazepam.

Initial understanding of the relationship between benzodiazepines and the GABA receptor was developed between 1975 and 1985 and based on pharmacological, electrophysiological, and radioactive ligand studies. In the last five years however a new level of understanding, and a vast potential for future improvement in therapy has been opened up by the purification, sequencing and cloning of the GABA/BZ receptor. Initial studies using purification by affinity chromatography and immunoprecipitation indicated that the receptor consisted of at least two subunits (designated α and β) (Sigel and Barnard, 1984). The purification of these subunits yielded fragments that allowed the design of synthetic oligodeoxyribonucleotide probes. These were used to screen cDNA libraries from bovine brain. DNA sequencing of hybridizing clones led to specification of the nucleotide and (deduced) amino acid sequences for the α and β subunits of the $GABA_A$ receptor (Schofield et al., 1987). These subunits each possess four hydrophilic membrane-spanning helices that are presumed to provide a central ion channel. Coexpression of the α and β subunit RNAs in Xenopus oocytes produces a functional receptor and ion channel. This shows the expected changes in Cl^- conductance in response to GABA or muscimol, with blockade by bicuculline or picrotoxin. Subsequently it was shown that a third subunit (designated the γ-subunit) is required for the functional expression of a receptor demonstrating the potentiating effect of benzodiazepines. Similar molecular biological studies have now been performed with GABA/BZ receptors from rat brain, and to a lesser extent, human brain.

GABA/BZ Receptor Heterogeneity

The earliest studies of ligand binding in brain membrane preparations (using [3H] diazepam or [3H]flunitrazepam) were interpreted as showing the existence of a single receptor site (Möhler and Okada, 1977; Squires and Braestrup, 1977). A separate acceptor at which diazepam and Ro 5–4864 (4'-chlorodiazepam) and PK 11195 (an isoquinoline carboxamide) bind, is referred to as the peripheral or ω benzodiazepine receptor. It is found in peripheral organs such as the kidney, but also in the brain where it occurs in astrocytes. Its functional role is not understood (it is not linked to a GABA receptor or to Cl^- conductance). A variety of molecules other than benzodiazepines have been shown to act on the benzodiazepine site in the GABA/BZ receptor complex. These include triazolopyridazines, such as CL 218872, and β-carbolines (see Sieghart, 1988). Studies with these non-benzodiazepine ligands led to the postulation of two BZ receptor subtypes (the BZ_1 and BZ_2 receptors) with differential regional expression in the brain (BZ_1 receptors predominating in the cerebellum but being equally expressed with BZ_2 receptors in the hippocampus).

Recent studies employing sequencing, cloning and transfection have greatly extended our understanding of the heterogeneity of the GABA/BZ receptor. It is now clear that there are as many as 6 different subtypes of the α-subunit and at least 3 forms of the β-subunit and 2 forms of the γ

subunit. Co-expression of an α, β, and γ subunit is required to give a functional unit showing the full pharmacology of the normal GABA/BZ receptor, although RNA corresponding to a single type of either the α or β subunit is sufficient for expression of a functional ion-channel responding to GABA and barbiturates (Levitan et al., 1988; Blair et al., 1988). This provides an enormous potential heterogeneity of GABA/BZ receptors as expression of the variants of the subunits appears to be under independent genetic control, and to show marked regional variation within the nervous system. Receptors with different variants of the α subunit show different sensitivity to GABA. The α subunit also determines whether the receptor shows the pharmacological properties of the BZ_1 or BZ_2 receptor. The α_1 variant confers the properties of the BZ_1 receptor (binding preferentially with CL 218872, 2-oxoquazepam and some β-carbolines), the α_2 and α_3 variants confer the properties of the BZ_2 receptor (Pritchett et al., 1989). Action at the BZ_1 receptor appears to be associated with greater anxiolytic action and less ataxia than action at the BZ_2 receptor. It is now possible to use transfected cells to study the physiology and pharmacology of all the possible combinations of subunit variants. This and the use of hybridization procedures to study the expression of these variants in the different regions of the brain opens up enormous possibilities both in the development of novel drugs and in the understanding of mechanisms involved in epilepsy (Table I).

Significance of GABA/BZ Receptor
Heterogeneity for Understanding the Aetiology of Epilepsy

It is possible that the occurrence of epilepsy is related to either inherited or acquired abnormalities in the GABA/BZ receptor system. With the use of tritiated ligands abnormalities have been demonstrated in the GABA/BZ receptor system in two genetically-determined syndromes of epilepsy in

TABLE I. GABA/BZ Receptor Heterogeneity

	BZ_1	BZ_2
Selective ligands	CL 218872 2-oxoquazepam β-CCM	
Regional distribution	Universal cerebellum (100%)	Hippocampus, striatum
Contribution to		
anxiolysis	++	+
ataxia	(+)	++
Subunit variants	α_1	α_2 α_3
	β_1 β_2 β_3	β_1 β_2 β_3
	γ_1 γ_2	γ_1 γ_2

Data from study in transfected cells (Pritchett et al., 1989).

rodents (see Horton this volume). In seizure-susceptible gerbils there is a reduction in [^{3}H]flunitrazepam binding in the midbrain (Olsen et al., 1985) and there is a similar deficiency in DBA/2 mice that show sound-induced seizures (Olsen et al., 1986). This is not, of course, proof that there is a primary defect in the GABA/BZ receptor system. There could for example be a primary defect in some other system (e.g., in kinases or other enzymes determining receptor and membrane properties) with a secondary change in the GABA/BZ receptor. Among the different forms of epilepsy occurring in man several are known to have a primary genetic aetiology (e.g.; *absence* seizures with 3 per sec spike and wave, juvenile myoclonic epilepsy), and many forms have some genetic factors contributing to their occurrence. The large number of genes involved in the expression of GABA/BZ receptors, and the likely subtlety of the functional effects of any abnormalities increases the possibility that some abnormalities of this system are contributory to epilepsy and provides encouragement to the search for such defects in patients with epilepsy.

Significance of GABA/BZ Receptor Heterogeneity for the Future Therapy of Epilepsy

The discovery of non-benzodiazepine compounds that interact with the benzodiazepine receptor led to the observation that these compounds frequently differed from the established anticonvulsant 1, 4-benzodiazepines in terms both of their anticonvulsant spectrum in animal models of epilepsy and their side-effect profile. Figure 1 shows the formulae of several β-carbolines shown to possess anticonvulsant activity in rodent and primate models of epilepsy (Meldrum et al., 1986; Turski et al., 1990). Table II provides a comparison of potencies of these β-carbolines and diazepam against sound-induced seizures in DBA/2 mice and photically-induced myoclonus in baboons (*Papio papio*). ZK 91296 and ZK 112119 (Abecarnil) have a very much weaker sedative and muscle relaxant action than diazepam in

TABLE II. Anticonvulsant β-carbolines Compared With Diazepam in 2 Models of Reflex Epilepsy

	DBA/2 mice ED$_{50}$ (mg/kg)	Papio papio ED$_{100}$ (mg/kg)
ZK 91296	0.50	1.0
ZK 93423	0.01	0.5
ZK 112119	0.33	0.1
Diazepam	0.0034	0.5

Values in DBA/2 mice are ED$_{50}$ for suppression of clonic or tonic seizures (with i.p. administration).
Values in *Papio papio* are ED$_{100}$ for suppression of photically-induced myoclonus (with i.v. administration).
(Chapman et al., 1987; Turski et al., 1990.)

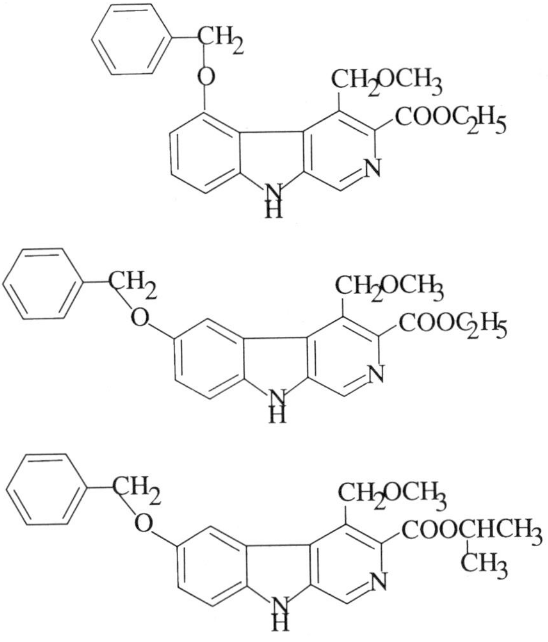

ZK 91296

ZK 93423

ZK 112119

Fig. 1. Structural formulae of 3 β-carbolines that are partial agonists at the benzodiazepine site on the GABA/BZ receptor. ZK 112119 is also known as Abecarnil.

both rodents and primates. In rats treated with etorphine, diazepam shows a muscle relaxant effect at 2 mg/kg but abecarnil requires 100 mg/kg.

It was possible to show (through antagonism with Flumazenil (Ro 15, 1788) and similar procedures) that the anticonvulsant and other effects of triazolopyridazines and β-carbolines are dependent on activity at the benzodiazepine receptor. Two explanations were put forward for the different effects of the benzodiazepines and β-carbolines. One concerned partial agonism. The weaker partial agonists were assumed to be very effective in audiogenic seizures in DBA/2 mice and in threshold pentylenetetrazol seizure but less effective in some other models and less prone to produce sedation and ataxia. The alternative explanation concerned BZ receptor subtypes, with a preferential action on BZ_1 receptors (as shown by CL 218872 and some β-carbolines) being associated with a relative lack of motor side effects. Although effects relating to partial agonism may exist, most evidence now supports the idea that differences in anticonvulsant and side effect profile between agents acting on the benzodiazepine receptor relate to different effects on receptor subtypes. The existing range of BZ-like drugs were derived by the testing of novel structural series in BZ receptor assays using brain membrane preparations, and then by evaluation in in vivo screens. Now it is possible to use transfected cells expressing subtypes of the GABA/BZ receptor (defined in terms of the variants of each sub-unit) to screen novel compounds. In this way it will be possible to perform very precise structure/activity studies in novel series and to optimize the spectrum of anticonvulsant and other effects.

GABA AGONISTS

Potent selective $GABA_A$ agonists such as muscimol and THIP are not suitable for trial as antiepileptic agents. In rodent models of epilepsy there is a very narrow margin between doses suppressing seizures and doses inducing severe side effects (including seizures). They facilitate the spike and wave discharges seen in the natural model of *absence* seizures in Wistar rats. In primates they induce diffuse myoclonus and facilitate spike and wave discharges and yet fail to suppress photically-induced myoclonus (Pedley et al., 1979; Meldrum and Horton, 1980). Agents that are selective $GABA_B$ agonists are also not suitable as antiepileptic agents. Baclofen also facilitates paroxysmal EEG discharges in primates. It was proposed that compounds that showed a weak $GABA_A$ and $GABA_B$ agonist activity (such as Progabide and SL 75102) might provide more satisfactory anticonvulsant activity (Lloyd et al., 1982). The exact mechanism for the anticonvulsant activity of Progabide has not been established and it remains an open question whether some limited forms of GABA agonist activity might be useful.

It would appear that supplying GABA agonists systemically so that their extracellular concentration is diffusely and chronically raised is unlikely to be a useful antiepileptic strategy. This is because there is relatively rapid accommodation to the inhibitory action of GABA at several central receptors (such as those in the hippocampus), and because the antiepileptic effect of activation of GABAergic interneurons is related to their specific function in the local circuitry with feedback inhibition becoming significant when there is an excessive local discharge. Diffuse chronic application of agonists fails to provide the temporal and spatial specification that is the key element in the normal inhibitory system. These disadvantages might not be so critical for compounds acting on the major GABAergic output pathways of the basal ganglia that modulate seizure expression and development. GABAergic agents injected focally into the substantia nigra pars reticulata or the entopeduncular nucleus in rodents suppress limbic seizures. Conceivably studies of receptor subtypes might lead to the development of drugs with selectivity for this site. Such drugs would need to leave unimpaired the subsequent GABAergic relay in the superior colliculus (where the effect of GABAergic activity on seizure threshold is reversed).

GABA given systemically does not penetrate the blood-brain barrier significantly, and high doses produce cardiovascular effects before they produce central effects. However three approaches have been developed that might provide central delivery of GABA. One is that of prodrugs that penetrate the blood-brain barrier and yield GABA. The principal compounds tested have been simple esters of GABA (benzoylGABA, pivaloylGABA and cetylGABA).

Anticonvulsant effects of these esters have been described in rodent models of epilepsy (Galzigna et al., 1978; Frey and Löscher 1980). GABA can also be given systemically in a liposome-entrapped form (i.e., sonicated with phosphatidylserine). Protection against isoniazid induced seizures has been reported following the i.p. injection of such a preparation (Loeb et al., 1986). The third delivery system utilizes a carrier containing a pyridine ring that participates in dihydro-pyridine-pyridinium redox reactions. GABAbenzyl ester is linked via an amide bond to the redox carrier which is lipophilic in its dihydropyridine form. In the brain this is oxidised to a charged quaternary complex that is trapped in the brain and hydrolysed to yield GABA. An anxiolytic action of this complex has been reported (Anderson et al., 1987).

It may also be practicable to deliver GABA or GABAergic agents focally into the brain. This has been shown to be effective in animal models of epilepsy. Several strategies are possible. The most direct is chronic infusion into an epileptic focus. In baboons with photosensitive epilepsy (*Papio papio* from Senegal) cortical discharges originate in the frontal cortex. Chronic focal infusion of GABA or benzodiazepines (by means of intracortical cannulae and osmotic minipumps) leads to suppression of photosensitive myoclonus) (Silva-Barrat et al., 1988). The alternative strategy is to provide chronic or intermittent focal infusion of GABAergic agents at strategic control points. The outputs of the basal ganglia (i.e., the globus pallidus or its analogue in rodents the entopeduncular nucleus and the substantia nigra, pars reticulata) exert a control over the development and expression of limbic seizures (Meldrum 1989). It is in principle possible to implant a cannula with a programmable pump operated by micro/computer that detects spike discharges such that injections would be given directly following the onset of focal seizure discharges. The most practical way of initiating such studies would be to implant guide cannulae and pumps at the same time as depth electrodes in the preliminary neurosurgical evaluation of patients being assessed for resective surgery for intractable seizures.

GABA-UPTAKE INHIBITORS

The effectiveness of GABA uptake inhibitors as anticonvulsants has been reviewed by Schousboe in Chapter 8. Orally active analogues of guvacine and nipecotic acid, such as SKF 100330A and NO 328 (Tiagabine) are potent anticonvulsants in several animal models (Yunger et al., 1984; Braestrup et al., 1990). Uncertainties concerning prospective clinical value relate to their side effects. In principle blocking reuptake should be less problematic than swamping receptors with GABA agonists, as the temporal relationship of enhanced GABAergic transmission to excessive local discharge should be preserved. However GABA reuptake into GABAergic terminals may play an important role in the metabolic economy of GABA and any uptake inhibitor that is itself taken up may displace GABA from stores and be released as a false transmitter. It is not yet clear whether the side effects reported with some of the GABA uptake inhibitors, that resemble those associated with muscimol (i.e., myoclonus and some types of confusional psychosis) are intrinsic to GABA uptake inhibition or can be avoided by an appropriate spectrum of inhibitory activity between glia and neurons. The most promising compound, tiagabine, is nevertheless under development by Novo and Abbott (Pierce et al., 1990). It is anticonvulsant in mice at 1 mg/kg i.p. In photosensitive baboons it produces only partial protection at 0.1 - 0.25 mg/kg i.p. but is associated with abnormal movements at 1 mg/kg. Phase I trials in man indicate that tiagabine is well tolerated in doses up to 10 mg daily.

GABA-TRANSAMINASE INHIBITORS

GABA transaminase inhibitors have been reviewed by Tunnicliff in Chapter 9. Reviews concerning vigabatrin are provided by Schechter (1986) and by Richens (1989). Vigabatrin is probably the most effective novel anticonvulsant agent to be introduced clinically since valproate. It is likely to have a significant role in the future management of "drug-resistant" epilepsy. The extent to which it will displace phenytoin or carbamazepine in the management of partial seizures cannot be predicted as it will depend on policy decisions by regulatory agencies and by marketing managers. There is no immediate prospect of Vigabatrin being replaced by an alternative GABA-transaminase inhibitor, but in the long-term this possibility cannot be excluded.

ALTERNATIVE STRATEGIES
RELATED TO GABA RECEPTOR MANIPULATION

The real goal of therapy in epilepsy should be an (instantaneous) curative procedure. This might be provided by some manipulation of the GABA receptor that permanently altered its sensitivity or distribution. One should also consider the possibility of preventive measures. In the developed world a high proportion of adult cases of epilepsy arise as a result of cerebral trauma (motor traffic accidents and interpersonal violence). Treatments influencing changes in GABA receptors in the post-traumatic period might prevent development of epilepsy. Genetic factors determining or influencing the development of epilepsy may also involve the GABA receptor. Therapeutic procedures would not necessarily involve replacement of defective genes in embryos. Modification of the local expression of genes determining GABA receptor subtype may be all that is required. Such procedures may be critical for the two other approaches, i.e., the curative treatment and the prevention of post-traumatic epilepsy. Thus discovering how to influence the expression of the various forms of the GABA/BZ receptor must be a major goal for those interested in the therapy and prevention of epilepsy.

REFERENCES

Anderson WR, Simpkins JW, Woodard PA, Winwood D, Stern WC, Bodor N: Anxiolytic activity of a brain delivery system for GABA. Psychopharmacology (Berl) 92:157–163, 1987.

Blair LAC, Levitan ES, Marshall J, Dionne VE, Barnard EA: Single subunits of the GABA$_A$ receptor form ion channels with properties of the native receptor. Science 242:577–579, 1988.

Braestrup C, Nielsen EB, Sonnewald U, Knutsen LJS, Andersen KE, Jansen JA, Frederiksen K, Andersen PH, Mortensen A, Suzdak PD: (R)-N-[4,4-Bis(3-Methyl-2-Thienyl)but-3-en-1-yl] Nipecotic acid binds with high affinity to the brain γ-aminobutyric acid uptake carrier. J Neurochem 54:639–647, 1990.

Chapman AG, De Sarro GB, Premachandra M, Meldrum BS: Bidirectional effects of β-carbolines in reflex epilepsy. Brain Res Bull 19:337–346, 1987.

Frey HH, Löscher W: Cetyl-GABA: effect on convulsant thresholds in mice and acute toxicity. Neuropharmacology 19:217–220, 1980.

Galzigna L, Garbin L, Bianchi M, Marxotto A: Properties of two derivatives of γ-aminobutyric acid (GABA) capable of abolishing cardiazol- and bicuculline-induced convulsions in the rat. Arch Int Pharmacodyn Ther 235:73–85, 1978.

Levitan ES, Schofield PR, Burt DR, Rhee LM, Wisden W, Köhler M, Fujita N, Rodriguez HF, Stephenson A, Darlison MG, Barnard EA, Seeburg PH: Structural and functional basis for GABA$_A$ receptor hetereogeneity. Nature 335:76–79, 1988.

Lloyd KG, Arbilla S, Beaumont K, Langer SZ, Bartholini G: γ-Aminobutyric acid (GABA) receptor stimulation. II Specificity of progabide (SL 76002) and SL 75102 for the GABA receptor. J Pharmacol Exp Ther 220:672–677, 1982.

Loeb C, Besio G, Mainardi P, Scotto P, Benassi E, Bo GP: Liposome-entrapped γ-aminobutyric acid inhibits isoniazid-induced epileptogenic activity in rats. Epilepsia 27:98–102, 1986.

Meldrum BS, Horton RW: Effects of the bicyclic GABA agonist, THIP, on myoclonic and seizure responses in mice and baboons with reflex epilepsy. Eur J Pharmacol 61: 231–237, 1980.

Meldrum BS, Kehr W, Stephens DN, Bhargava AS: ZK 91296 and ZK 93423: β-carbolines with partial or full agonist activity at benzodiazepine receptors. In Meldrum BS, Porter RJ (eds): "New Anticonvulsant Drugs," London: John Libbey, 1986, pp.303–308.

Meldrum BS: GABAergic mechanisms in the pathogenesis and treatment of epilepsy. Br J Clin Pharmacol 27:3S–11S, 1989.

Möhler H, Okada T: Demonstration of benzodiazepine receptors in the central nervous system. Science 198:849–851, 1977.

Olsen RW, Wamsley JK, McCabe RT, Lee RJ, Lomax P: Benzodiazepine/γ-aminoburyric acid receptor deficit in the midbrain of the seizure-susceptible gerbil. Proc Nat Acad Sci USA 82:6701–6705, 1985.

Olsen RW, Wamsley JK, McCabe RT, Lee RJ, Lomax P, Seyfried TN: Midbrain GABA receptor deficit in genetic animal models of epilepsy. In Nistico G, Morselli PL, Lloyd KG, Fariello RG, Engel J (eds): "Neurotransmitters, Seizures and Epilepsy III." New York: Raven Press, 1986, pp 279–291.

Pedley TA, Horton RW, Meldrum BS: Electroencephalographic and behavioural effects of a GABA agonist (muscimol) on photosensitive epilepsy in the baboon, Papio papio. Epilepsia 20:409–416, 1979.

Pierce MW, Suzdak PD, Gustavson LE, Mengel HB, Mckelvy JF, Mant T: Tiagabine. In Perucca E, Pisani F, Richens A (eds): "The New Antiepileptic Drugs". Amsterdam: Elsevier, 1990.

Pritchett DB, Luddens H, Seeburg PH, Type I and Type II GABA$_A$-benzodiazepine receptors produced in transfected cells. Science 245:1389–1392, 1989.

Richens A: Potential antiepileptic drugs: Vigabatrin. In Levy RH, Dreifuss FE, Mattson RH, Meldrum B, Penry JK (eds): "Antiepileptic Drugs." New York: Raven Press, 1989, pp 936–946.

Schechter PJ: Vigabatrin. In: Meldrum BS, Porter RJ (eds): "New Anticonvulsant Drugs." London: John Libbey, 1986, pp 265–275.

Schofield PR, Darlison MG, Fujita N, Burt DR, Stephenson FA, Rodriguez H, Rhee LM, Ramachandran J, Reale V, Glencorse TA, Seeburg PH, Barnard EA: Sequence and functional expression of the $GABA_A$ receptor shows a ligand-gated receptor superfamily. Nature 328:221–227, 1987.

Sieghart W: Multiple benzodiazepine binding sites. In R.F. Squires (ed): "GABA and Benzodiazepine Receptors," Vol. 2. Boca Raton: CRC Press, 1988, pp 1–14.

Sigel E and Barnard EA (1984): A γ-aminobutyric acid/benzodiazepine receptor complex from bovine cerebral cortex. J Biol Chem 259:7219–7223.

Silva-Barrat C, Brailowsky S, Riche D, Menini Ch: Anticonvulsant effect of localized chronic infusions of GABA in cortical and reticular structures of baboons. Exp Neurol 101:418–427, 1988.

Squires RF and Braestrup C: Benzodiazepine receptors in rat brain. Nature 266:732–734, 1977.

Turski L, Stephens DN, Jensen LH, Petersen EN, Meldrum BS, Patel S, Bondo Hansen J, Löscher W, Schneider HH, Schmiechen R: Anticonvulsant action of the β-Carboline abecarnil—Studies in rodents and baboon, Papio papio. J Pharmacol Exp Ther, 1990.

Yunger LM, Fowler PJ, Zarevics P, Setler PE: Novel inhibitors of γ-aminobutyric acid (GABA) uptake: Anticonvulsant actions in rats and mice. J Pharmacol Exp Ther 228:109–115, 1984.

Index